Withdrawn
OCT 2 1 2024
UNBC Library

Tie Hackers to Timber Harvesters

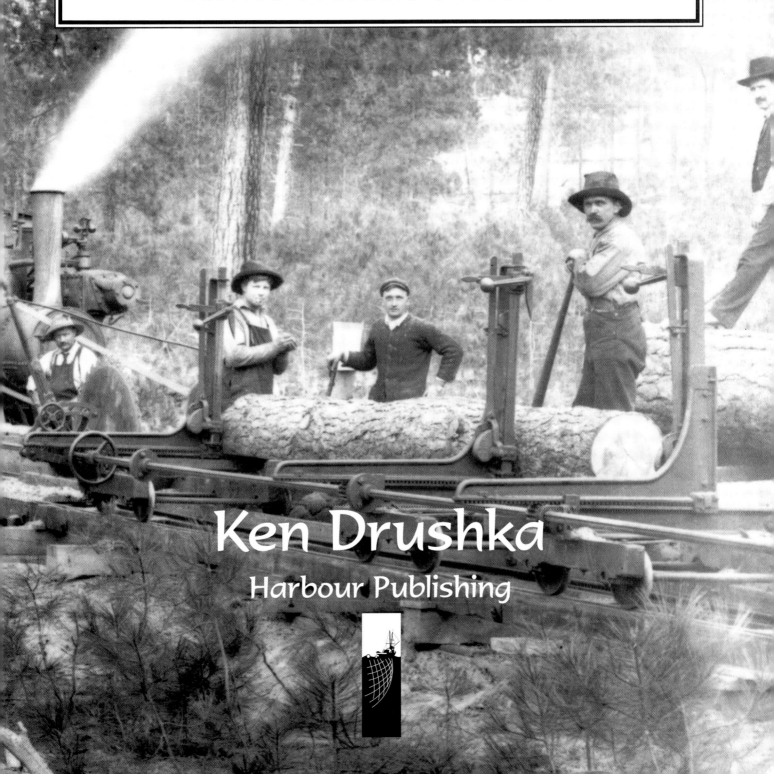

Tie Hackers to Timber Harvesters

THE HISTORY OF LOGGING IN BRITISH COLUMBIA'S INTERIOR

Ken Drushka

Harbour Publishing

Copyright © 1998, Ken Drushka.

No part of this publication may be reproduced, stored in a retrieval system or transmitted, in any form or by any means, without prior permission of the publisher or, in case of photocopying or other reprographic copying, a licence from CANCOPY (Canadian Reprography Collective), 214 King Street West, Toronto, Ontario, M5H 3S6.

Harbour Publishing,
PO Box 219,
Madeira Park, BC V0N 2H0

Cover design, page design and composition by Roger Handling, Terra Firma Digital Arts

Harbour Publishing acknowledges the financial support of the Government of Canada through the Book Publishing Industry Development Program and the Province of British Columbia through the British Columbia Arts Council, for its publishing activities.

The Canada Council for the Arts since 1957 | Le Conseil des Arts du Canada depuis 1957

Printed in Canada.

Canadian Cataloguing in Publication Data

Drushka, Ken
    Tie hackers to timber harvesters

    Includes bibliographical references and index.
    ISBN 1-55017-189-5

    1. Logging—British Columbia—History. I. Title.
SD538.3.C2D78 1998   634.9'8'09711   C98-910984-4

Photograph credits: BCARS = BC Archives & Records Service; BVM = Bulkley Valley Museum; CFI = Crestbrook Forest Industries; CFP = Canadian Forest Products (Canfor) collection; EDMS = Enderby & District Museum Society; FFG = Fraser–Fort George Regional Museum; FSM = Fort Steele Heritage Town Archives; GDHS = Golden and District Historical Society; KLM = Kelowna Museum; KMA = Nelson Museum/Kootenay Museum Association and Historical Society; KM = Kamloops Museum; MDMS = Mackenzie & District Museum Society; NAC = National Archives of Canada; NVMA = Nicola Valley Museum Archives Association; PGL = Prince George Public Library; QDMA = Quesnel & District Museum & Archives; SG = Slocan Group collection; TI = Tolko Industries collection; TPL = The Pas Lumber collection; UBC = University of British Columbia Library, Special Collections and University Archives; USNR = USNR Kockums CanCar; WFT = West Fraser Timber collection; WMCR = Whyte Museum of the Canadian Rockies.

Frontispieces: Clearing a jam during an early river drive on the Shuswap River. EDMS;
A portable sawmill at work in the woods, 1906. G.H.E. Hudson photograph, courtesy Kelowna Museum.

CONTENTS

	Introduction	7
Chapter 1	**The Early Days** The 1850s, '60s and '70s	11
Chapter 2	**By Rail & By River** The 1880s and 1890s	21
Chapter 3	**Growing Pains** The Early 1900s	39
Chapter 4	**Boom Years** The 1910s	61
Chapter 5	**Supply, Demand & Depression** The 1920s	81
Chapter 6	**Hacking & Hauling** The 1930s	103
Chapter 7	**Growing & Prospering** The 1940s	121
Chapter 8	**Brokers & Bush Mills** The 1950s	147
Chapter 9	**Bigger Plans, Bigger Money** The 1960s	171
Chapter 10	**Technological Transformation** The 1970s	193
Chapter 11	**New Challenges** The 1980s and 1990s	211
	Oral history sources	237
	Index	238

Introduction

A record one-horse load of eighty-one ties, at Weatherhead's camp near Yahk, 1925. BCARS 56355

The idea of writing a history of the Interior British Columbia forest industry first arose a few years ago when I began a book on coastal logging history. At that time I had considered, briefly, attempting a book that would document the evolution of the industry throughout the entire province. It quickly became clear this was probably an impossible task. The subject was simply too big for one book, and the coastal and Interior industries are too different to include under one cover.

The Interior of BC is an extremely diverse place. It consists of several distinct regions, each with its own pattern of development. Until relatively recent times, these regions were isolated from each other and consequently developed their own characters. Forest workers in some regions hardly know of the existence of other regions, let alone how logging and sawmilling have evolved there. The differences between climate and terrain in the Okanagan and the Bulkley Valley, for instance, are greater than those between most nations, and the industrial development of Prince George and Cranbrook followed completely different paths, decades apart.

Compared to the Coast, the Interior industry has been dominated less by large, integrated corporations than by local entrepreneurs and family-owned companies. Until recently, there have been few Interior companies that operate in more than one part of the province, so a diverse cast of characters appears in this account. Thousands of people and hundreds of companies have played significant roles in the evolution of the industry. For every name mentioned in this book there are scores—probably hundreds—more whose contributions go unrecorded.

Before I began, the variety and diversity did seem intimidating. More than one knowledgeable person suggested it would not be possible to obtain a coherent overview of the entire Interior's forest history. There were simply too many people, too many companies, and too much variation in their evolution. Those people were right, but the fact that no one had ever attempted the task gave me a compelling reason to take a stab at it. It was fresh territory.

When I began discussing the idea with a few of the Interior forest industry's leading figures, I received an enthusiastic response. There was a widespread belief that such a history was long overdue. The heads of several companies quickly agreed to finance the undertaking,

The Ross-Saskatoon Company railway logging camp near Waldo. The company operated from 1912 to 1932. The locomotive is a Shay and is running on a narrow gauge railway. This is a typical railway logging camp of the time.

with no suggestion they should have any say in the end result. It was a unique opportunity to travel to every region of the Interior and talk with a wide range of people with deep roots in the forest industry and in their communities. The assistance and advice I obtained from these scores of people, many of whom went far out of their way to be helpful, turned a difficult task into an exciting and pleasurable undertaking.

This book is not an exhaustive history of the Interior forest industry: that story would require several volumes. The history of the industry in every region—not to mention in dozens of individual valleys—is worth a book in itself. In fact, a few of these have been written, and I found them invaluable in my research. Without them, the job would have been a lot harder. Ken Bernsohn's *Cutting Up The North* is a superb account of how the industry evolved around Prince George. Mike Halleran's *Loggers and Lumbermen* provides a precise description of the corporate evolution of the industry in the southern Interior. Jack Mould's *Stump Farms and Broadaxes* is a wonderful story about the early forest industry of the Bulkley Valley, as is Verdun Casselman's *Ties to Water*, the history of the forest industry along the Bull River in southeastern BC. Dozens more books like these are yet to be written. What I have attempted in the following pages is to provide an Interior-wide overview that sets a historical context for any number of more dramatically detailed accounts.

For me, the most painful part of the undertaking has been to see at close range the sorry state to which the Interior forest industry, along with that on the Coast, has been reduced. The research and writing have been done at a time when the fortunes of the industry are in decline. Amidst all the diversity, the one common conviction I encountered is that this situation need not exist. Everyone I talked to believes that a vastly enhanced industry is possible, that we could be doing a lot better than we are at providing the people of BC with economic benefits without in any way adversely affecting our forest ecology.

Central to every pivotal work of philosophy is the connection between past and future. I believe that if we do not understand where we've been, we'll never figure out where we're going. My hope is that this book will provide some assistance to those whose task it is to restore the health and vitality of the most important economic endeavour in the BC Interior.

A lot of people made this book possible. Foremost among them is Jake Kerr, whose enthusiasm from the outset provided the resources needed to get the work done. Jim Collins has, once again, dispensed information and advice that I could not have done without.

I would like to acknowledge with thanks and appreciation the financial support of Canadian Forest Products, Weldwood of Canada, Riverside Forest Products, The Pas Lumber, Ainsworth Lumber, Dunkley Lumber, West Fraser Timber, Tolko Industries, Slocan Group and Lignum. I also thank the dozens of people in these companies who helped with information, memories, ideas and photographs.

Scattered throughout the book are excerpts of discussions I had with several dozen veterans of the industry, and I thank them for their time and patience. I am grateful for similar assistance from Henry Novak, Gordon Steele, Frank Druegel, John Brink, Roger Getz, Jim Rustad, Hans Scholz, John Dahl, John Harding, Harold Patenaude, Bill Kordyban, Ed Peers, John Krauchi, Dirk Septer, Clive Stangoe, Ken Bernsohn, Mike Halleran and my old friends Martin Lynch and Jarl Sundve. George Killy deserves special thanks for aiming me in the right direction, as does Wendi Struthers for attending diligently to the details.

There is a category of people without whom our knowledge of our history would be much diminished: the professional historians, librarians and archivists who work in various institutions throughout the province. They include George Brandak (Special Collections, University of British Columbia Library), Richard Rajala, Jane Turner (University of Victoria), Bill Quackenbush (Barkerville), Derryll White (Fort Steele), Shawn Lamb (Nelson Museum), Colleen Torrence (Golden Museum), Ron Welwood (Selkirk College), Lillian Weedmark (Smithers Museum), Jude Cooper (Fraser–Fort George Museum), Joan Cowan (Enderby Museum), Elizabeth Duckworth (Kamloops Museum), Lee Fraser (Chase Museum), Randy Manuel (Penticton Museum), Ruth Stubbs (Quesnel Museum), Karen Little (Mackenzie Museum), Ursula Surtees (Kelowna Museum) and Barbara Bell (Vernon Museum).

Finally, a thank you to Mary Schendlinger for her meticulous editing, Roger Handling for a great design and Peter Robson and Derek Fairbridge for keeping it all together.

CHAPTER 1

The Early Days

The 1850s, '60s & '70s

When the first Europeans arrived in the interior of British Columbia in the late eighteenth century, they found a series of forests which had been growing for ten to twelve thousand years. These were the forests which evolved after the retreat of the last glaciers. As the ice melted, various plant and animal species migrated north from their southern refuges and repopulated the scoured land. The wide variety of geographical conditions found in BC resulted in a great variety of forest types, perhaps the most diverse in the world.

The dominant feature of the BC landscape is a series of mountain ranges running generally northwest to southeast. The Interior is defined on the west by the Coast Mountains running from the province's southern border on the 49th parallel to the Alaska border on the Portland Canal north of Prince Rupert. The eastern Interior is partially demarcated by the towering ramparts of the Rocky Mountains, from the US border in the south to a point east of Prince George. Here, the provincial boundary makers abandoned geographical considerations and pushed the border due north to the 60th parallel, encompassing a large triangle of northern plains lying east of the Rockies.

At one time in the distant past, according to geologists, the area now occupied by the Rocky Mountains once constituted the western shore of northern North America. The land now known as British Columbia was slowly pushed up out of the Pacific Ocean by the gradual drift of continental plates. Over many hundreds of millions of years, unimaginable forces drove the ancient shoreline thousands of metres above sea level and shaped this land into a series of mountain ranges separated by valleys, some of them broad, others steep and narrow. Through the centre of the province, from the US to the Yukon borders, a vast plateau was formed. Apart from the province's northeastern plains, which consist of a deep layer of undisturbed sediments covering the original core of the North American continent, the BC Interior is a relatively young land.

During various glacial periods, when the Interior was covered with ice, plants and animals migrated to ice-free areas, and returned when temperatures warmed and the ice melted. The factor determining which species populated the various parts of this landscape was climate. The wetter climate on the western slopes of the highest eastern mountain ranges, for example, favoured coastal tree species such as Douglas fir, western red cedar and western hemlock, while in the drier southern plateau, the pines competed most successfully for living space. Farther north, where a major climatic influence is cold Arctic winter air, hardier species such as spruce dominated.

In an attempt to understand this geographical diversity, scientists have devised various classification systems, including the Ecoprovince Map (see page 6). Except for the Southern Interior Mountain ecoprovince, which includes a dry region in its southeastern portion, and moist conditions elsewhere, this map pretty well coincides with Interior forest regions. Within each of these ecoprovinces the forests are similar and the forest industries have experienced the same historical development.

In broad strokes, this is the Interior forest. It takes up 82 percent of the province's 366,000 square miles (952,000 km²), stretching 760 miles (1216 km) north from the US border to the Yukon, and 400 miles (640 km) east from the Coast Mountains to the Rockies, and it contains an immense variety of timber types and terrain. Because of this natural diversity, and the fact that many areas were isolated from neighbouring regions for many decades before transportation networks evolved, the character of each area and its forest industry evolved separately and distinctly.

Pages 10–11: Falling crew with a big western red cedar during right-of-way clearance for the Big Bend highway near Donald, 1929. BCARS 73801

Early fallers, Bill (left) and Eli Johnston, use springboards to get above a tree's flared butt near Mabel Lake, 1898.

The forests in much of the Interior are unique, not only in BC but also in Canada. Although the northern boreal forests in the north central region are part of, and share some features with, the great northern boreal forest stretching from Alaska to Labrador, the southern half of the Interior forest is quite distinct. In most respects it is markedly different from BC coastal forests, which are characterized by very large trees. But while Interior trees are usually smaller, there are exceptions. For example, a century ago there were extensive stands of very large western red cedars in the Interior wet belt, especially around Revelstoke; some individual trees were 8 to 10 feet (2.4–3 m) in diameter. In the southern portion of the east Kootenays and in parts of the Okanagan, ponderosa pines of 5 to 6 feet (1.5–2 m) in diameter were common. In the central Interior and throughout the northern half of the province, the trees—primarily spruce and lodgepole pine—were generally much smaller, rarely growing to more than 2 feet (.6 m) in diameter.

By early 1858, this vast area had been barely penetrated by Europeans. The Hudson's Bay Company operated a few scattered trading posts throughout the Interior and the only coastal settlement of any consequence—Victoria, with a population of 250–300 people—was a sleepy colonial outpost. On a Sunday morning in late April, all that changed when the US sidewheeler *Commodore* unloaded 450 gold seekers onto the settlement's small wharf.

At this point in history, North America was in a state of turmoil. The United States was deeply divided and on the verge of civil war, but was also in an expansionist mood: its

Peeling mine props near Wardner, early twentieth century. FSM F.5.5-148

population swelled with immigrants and a huge westward migration threatened to head north into British North America. Meanwhile, the idea of Canadian confederation was afoot. Various expeditions were launched from Toronto and Montreal to investigate the feasibility of tying together the vast territory between Vancouver Island and the Great Lakes and forming one nation.

For a decade, tens of thousands of prospectors, speculators and adventurers of every description from practically every country in the world were caught up in a frenzied search for gold that began in California, then spread throughout what we know today as the Pacific Northwest. The first to arrive were drifters, vagrants and assorted hustlers, but gold fever soon infected more stable and respectable people, who quit their jobs and sold or abandoned their farms and businesses to join the obsessive search for a fast fortune.

In 1856 word reached the coast that gold had been found at the junction of the Pend d'Oreille and Columbia rivers, just below the newly established international border at the 49th parallel. Gold seekers pouring into that area ignored the border, pushing up the Columbia and overland into the valleys of the Thompson River. The following year, 1,400 ounces of gold was recovered around the junction of the Thompson and Fraser rivers. Over the winter this news circulated in San Francisco, and one day the following spring, just as the inhabitants of Victoria were emerging from church, an overloaded ship steamed into the harbour and unloaded a mob of would-be miners. Most members of this motley horde—described by one prominent Victorian as "an indescribable array of Polish Jews, Italian fishermen,

French cooks, jobbers, speculators of every kind, land agents, auctioneers, hangers on at auctions, bummers, bankrupts and brokers of every description"—found their way to the mouth of the Fraser River and up to the sand bars where the Fraser valley narrows into the canyon near Hope.

They obtained encouraging amounts of gold from the sand bars and, after the spring freshet subsided, were astounded to discover fresh deposits of fine gold. The news spread, and some 25,000 to 30,000 prospectors flocked to the Fraser. Over the next couple of years they pushed farther up the river, then inland. In the summer of 1860 a major deposit was discovered at Antler, deep in the heart of the Cariboo mountains. This touched off the Cariboo gold rush which culminated in the development of settlements at Barkerville and Quesnel.

> I will speak first of the effects of deforestation upon the earth. The orthodox history records that the great Creator planted a garden [of trees] near the headwaters of the Euphrates, and therein put the newly created man, charging him to prune and dress the trees, to take care of them, and to leave some of them severely alone. When man disobeyed the injunction he was thrust out.
>
> An alternative and modern view is that the branch of the family of Catarrhine apes from which we are descended developed opposable thumbs, and thereby became qualified to tear the branches from trees and use them as clubs, in the neighbourhood of the Euphrates valley. Both accounts agree that man began the wanton destruction of trees that has characterized him in central Asia, and the evidence is that he has progressed from that point in his career of devastation to the summit of the Rocky Mountains of America, leaving a blasted and desolate world behind him.
>
> We who live on the western slope of the Rockies are waiting now for our bankers to say our cheques will be good, to start up our logging and tie camps, and our sawmills, and finish (with the exception of a tract in Siberia, where the regulations are strenuous) what is left of useful natural timber on the face of the earth.
>
> —G.O. Buchanan, former owner of Buchanan Lumber at Kaslo, and an early member of the Canadian Forestry Association.

At approximately the same time, gold fever continued to lure miners into the southeastern Interior. In 1864 a major find was made at Wildhorse Creek, where it flows into the Kootenay River at the site of Fort Steele. Hundreds of miners poured into the area from Idaho and northeastern Washington, and roads and trails were built to accommodate them. Reports of gold finds at other Interior locations followed.

In one form or another the gold rush persisted for forty years or more, concluding with the final mad dash into northwestern BC and the Klondike in 1898. It was an obsessive, explosive phenomenon, an invasion of a land that had been inhabited only by generations of aboriginal people and a few fur traders. The gold rush left in its wake numerous settlements, occupied by immigrants who had the wherewithal and the tenacity to treat mining as a long-term proposition, by entrepreneurs whose business interests lay in other directions, and by people who liked the rugged beauty of this new place and decided to settle.

At this time BC was essentially in a preindustrial condition, except for steam-driven boats and a few small sawmills. Steamships carried most of the first rush of miners up the Fraser River to Hope, and in July 1858 during the first year of the rush, the sternwheeler *Umatilla* made the first ascent from Hope to Yale, a five-hour voyage against a strong current, urged on by several hundred passengers, cheering and firing their guns in the air. Five years later, the 110-foot-long *Enterprise* was built at Alexandria, above the Fraser Canyon. It was made of hand-sawn lumber and powered with a boiler and steam engine hauled 300 miles (480 km) overland by mules. This vessel, the first steam-powered machine to operate in the Interior, plied the Fraser for eight years, usually between Soda Creek (the southernmost navigable point before the canyon) and Quesnel. But in 1871, when gold was discovered in the Omineca country, the *Enterprise* was taken on an astounding voyage up the Fraser to Fort George, then up the Nechako River to the Stuart River and through that system to the head of Takla Lake.

The first sawmill to operate on the BC mainland was built at Yale, during the opening summer of the Fraser gold rush, by Land, Fleming and Company. It sold rough lumber for $125 a thousand board feet to the two thousand miners preparing to spend their first winter at Yale. Compared to today's spruce, pine and fir studs, which sell for $250 a thousand board feet, this was not cheap wood. Within a few years the boom had moved beyond Yale. In 1863 the mill was sold to another local operator who moved it closer to town, at which point it produced 6,000 board feet a day. Today, some modern Interior mills cut nearly a million board feet a day.

There is no record of what type of mill was built for this pioneering lumber business at Yale. Throughout the gold rush era, even beyond the turn of the century, a large portion of the lumber produced was cut by hand. This primitive method, which likely originated in Europe centuries earlier, was known in North America as whipsawing, or pit sawing. The saw was 6 to 8 feet (2–2.5 m) long and had a handle on each end. A log was rolled so that it straddled a pit, in which one of the two sawyers stood. He pulled the saw down and into the log on the cutting stroke; then his partner, who stood on the log, pulled the saw back up. Judging from the few photographs taken of these operations, pits were rarely used in BC. Instead, a frame was constructed onto which the log was raised, and the sawyer stood under this frame.

Enormous volumes of lumber were cut by this laborious method. Numerous steamboats were built in the Interior with whipsawn lumber, as were the first buildings in many mining towns. The construction of sluice boxes and flumes, in which gravel was washed in early placer operations, required a great volume of lumber; much of it was whipsawn.

In fact, the earliest powered sawmills were little more than mechanized whipsaws, in which human muscle power was replaced with the power harnessed by a water wheel. A dam constructed on a small stream provided a flow of water which was directed over, or occasionally

Whipsawing lumber at Bennett, near the Yukon border, 1907.
BCARS 77050

Ithiel Nason's water-powered sawmill at Richfield, near Barkerville, 1865.
BCARS 95331

under, a wooden paddle wheel. The wheel turned a shaft to which an eccentric was attached, and the eccentric raised and lowered the saw, which was fixed in a frame. The device was developed to the point where a series of saw blades were mounted within the frame or sash, so that several boards could be cut in one pass of the log through the saw.

Some of the earliest known water-powered mills were used in the fourth century, in what is now Germany. Their use spread throughout Europe (except in England, where pit sawyers destroyed sawmills until sometime in the 1760s in order to protect their jobs) and they were probably introduced to Canada in the seventeenth century, by the first Acadian settlers. The ability to construct a simple water-powered mill was a common skill by the late 1850s, when European settlement of the BC Interior began. One of the great virtues of this technology was that a water-driven mill could be built fairly easily in a remote location with local materials. A skilled millwright needed to import only a saw blade; the rest of the operation could be built from whipsawn or hand-hewn lumber and timbers.

The Land, Fleming mill may also have been steam powered. The first North American steam-powered sawmills were built on the Mississippi River in 1804, and within twenty-five or thirty years, foundries in eastern Canada and the US were producing steam engines designed to drive sash mills. By the 1850s, large circular rip saws with replaceable teeth had evolved and steam-powered circular saws began to appear throughout the industrialized world. Because they were heavy machines, these mills were slow to find their way into less accessible areas. Steamships began serving Yale in the spring of 1858, so it is possible Land and Fleming had a steam-powered mill shipped up the river and in operation by late summer.

The next mill to open in the Interior was a water-powered mill built by R.P. Baylor at Antler, during the height of gold fever in 1860. By the following year, Antler had sixty to seventy houses built of sawn lumber and was the largest settlement on the mainland. A year after that, with new gold discoveries concentrated around Barkerville, Antler was practically abandoned. Baylor moved his mill to Richfield, closer to Barkerville, and built a second water-powered mill.

At about the same time, G.B. Wright and Jerome Harper built what was probably the first steam-powered mill in the Interior, at Quesnel. It was built specifically to cut lumber for construction of the steamship *Enterprise*. When that task was completed, the mill produced lumber for construction of buildings, mining flumes, bridges, mine props and other uses.

In 1866 this mill was purchased by Ithiel Nason and moved to Richfield where, it was advertised, it cut 1,000 board feet an hour. Nason was typical of the more stable businessmen who moved into the Cariboo in the wake of the gold rush. He was born into a sawmilling family in Maine in 1840 and, after receiving a thorough education, joined the California gold rush, where he worked in a sawmill. He then followed the gold trail north, and ended up managing a sawmill at Langley before travelling to the Cariboo. In 1863 Nason took over a water-powered mill at Richfield—possibly the old Baylor mill—and three years later added the Quesnel steam mill. A few years after that, he built another mill at Quesnel to provide lumber for a bridge over the Quesnel River. He built a large and prosperous lumber business, became one of the region's most prominent citizens and served as the Cariboo MLA for two sessions.

> Loggers may be as hard-boiled as I am told, but I don't know why they shouldn't be credited with the virtues of good citizenship, public spirit and pride of race. If uninformed persons sometimes call you vandals it is simply because they are too blind to see that it is your job to convert our greatest natural resource to the use of man.
> —Fred Mulholland, who was a senior forester in the BC Forest Service until he resigned to become Comox Logging & Railway's first chief forester.

As settlement moved farther into the Interior, Quesnel developed as a distribution centre. A good supply of timber was available in the Cariboo mountains to the east, and logs could be floated down the Quesnel River, so several mills came and went through the 1870s. One of the most permanent was built by another pillar of the local business community, James Reid. A retail merchant who had come to the Cariboo from Quebec, Reid was a longtime sawmill owner who eventually sat in the Canadian Senate.

Elsewhere in BC, the gold rush was such a short-lived phenomenon that it provided little incentive for permanent settlement. For example, the strike at Wildhorse on the Kootenay River in 1864 lasted only a year, and a sawmill erected there also lasted only a year, by which time the miners had moved north up the Columbia valley to another brief boom at Big Bend.

However, a different kind of development was occurring at Kamloops, at the junction of the North and South Thompson rivers. Kamloops had been an important trading post since early in the century, and the various buildings and other structures erected by the Hudson's Bay Company were built of hewn timber logged up the North Thompson River and floated down to the fort. The HBC contracted with a man named Isaac McQueen to whipsaw lumber for construction of new buildings in 1862. Three years later, McQueen was hired to whipsaw lumber for an HBC steamboat, the *Marten*, built at Chase from timber cut along the shores of Shuswap Lake.

The first sawmill in the area was built in 1868, at Tranquille on the north shore of Kamloops Lake, by one of the *Marten*'s shipwrights, James McIntosh, and his neighbour William Fortune. This was a water-powered mill using water flumed from the Tranquille River. It may have been the first operation in BC to combine a grist mill, for grinding flour, with a sawmill. These combined mills, which were common in eastern Canada and the US earlier in the century, were designed on the principle that most of the effort of building a water-powered mill went into the construction of a dam, water wheel and flumeworks. Once this task was completed, it was relatively simple to install a grist mill, a sawmill or both.

During the 1870s the ranching and farming industries around Kamloops grew rapidly and local demand for lumber increased accordingly, leading to the establishment of other small mills. James Jamieson built his sawmill 12 miles (19 km) up the North Thompson and for ten

James Reid's sawmill, with the steamship **Charlotte***, Quesnel, 1897.*
QDMA P89.61.1

years rafted lumber to customers in town. Fortune and McIntosh ended their partnership and went on to form other partnerships in new mill enterprises. Fortune built the steamship *Lady Dufferin* to tow logs to his new mill. McIntosh became a partner in the much larger, steam-driven Shuswap Milling Company, which ground fifty barrels of flour as well as cutting 15,000 board feet of lumber a day on the present site of Kamloops' Riverside Park.

Another ranching centre was evolving south of Kamloops in the Nicola Valley in the early 1870s, and in 1871 George Fensom, a rancher, built a water-powered sawmill to provide his neighbours and himself with lumber, and an attached grist mill to grind flour. Six years later, Albert House walked into the valley over the Coquihalla Trail and, after a brief survey of the area, ordered a portable steam mill from Toronto which he operated at various locations in the valley before moving it to Princeton in 1899. This was the first full-time sawmill operation in the Merritt–Princeton area.

The invasion of American gold seekers in 1870–71 prompted the colony of British Columbia to decide to join Canadian confederation. In itself this event was of no great significance to most of the Interior, but one of the conditions of BC's membership—the construction of a transcontinental railway—was. The prospect of the railway stirred a lot of enthusiasm throughout the Interior for a couple of years. Then, in 1873, financial panic swept the continent. Combined with a political scandal in Ottawa over bribes paid to the government in exchange for railway-building contracts, the panic caused a recession and widespread despair about the future of the BC Interior. The gold rush subsided, development slowed and there was little growth of the Interior economy until after 1880, when a new railway contract was signed and construction began. The stage was set for a decades-long period of settlement and development, much of it dependent upon the health of a forest-based industry.

CHAPTER 2

By Rail & By River

The 1880s & 1890s

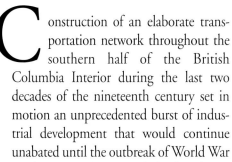

Construction of an elaborate transportation network throughout the southern half of the British Columbia Interior during the last two decades of the nineteenth century set in motion an unprecedented burst of industrial development that would continue unabated until the outbreak of World War I. This network consisted chiefly of main-line and branch-line railways and a system of steamships plying navigable lakes and rivers. Even by current standards the building of this system was an impressive accomplishment; under conditions prevalent in the early 1880s the achievements were astounding. To a great extent this entire endeavour relied upon the rapid development of forest industries—logging and sawmilling.

Construction of the Canadian Pacific Railway's main line, from the point where it entered BC at the top of Kicking Horse Pass, near Field, to Port Moody on Burrard Inlet, was in itself a mammoth undertaking that required vast volumes of timber. Except for a short section of the line that ran through the Fraser Canyon, roughly following the Cariboo Wagon Trail, there were no transportation facilities in place to aid in construction of the railway. All the building materials had to be brought in from the rear, over the newly built tracks. These included 3,000 ties for every mile of track, and millions of board feet of lumber for bridges, trestles, construction camps, stations, warehouses and so on. There was no established industry to provide these materials along most of the line. Loggers, who were in short supply, had to be hired to clear a right of way and then to provide logs for a series of mills, most of which were temporary.

Construction of the CPR line began in the spring of 1880 at Emory, a few miles south of Yale, on the basis of a contract between the Canadian government and an experienced US contractor, Andrew Onderdonk. This phase of the work began even before the CPR was formed, to placate the colonial administration of BC, which had threatened to secede from confederation because of delays in the start of railway construction.

There was intense pressure on Onderdonk to get the job done quickly and cheaply. Consequently most bridges, trestles, retaining walls and other structures were built of wood and later replaced by steel or masonry.

One of Onderdonk's first actions was to award a series of

tie-cutting contracts in the Fraser Valley and along the wagon trail at various points where suitable timber was available. He also set up permanent sawmills at Yale and Emory, and temporary mills at some of the major construction sites and timber stands along the route. To man these operations, he brought in experienced loggers and millworkers from California, Washington and Oregon, adding new capabilities to the province's labour force.

Very little mechanized equipment was available, particularly in the early stages of the project, so most construction work along the railway was done by hand. Bridges were built without the use of cranes or hoists; ties and rails were carried and spiked into place by human effort, much of it produced by workers imported from China. As construction proceeded beyond Yale, small locomotives were brought in to haul materials to the end of steel. The second of these, known as Curly, later became well known as the first logging locomotive in BC. Built at Union Iron Works in San Francisco in 1869, it was used there by Onderdonk before he brought it to BC. After the CPR job was finished, Curly was sold to Royal City Planing Mills and used to haul logs in Surrey, and eventually found its way to a well-earned retirement at the Burnaby Village Museum. But before that, Curly and the other locomotives hauled logs to the construction mills along the line—the first railway logging operations in BC.

While Onderdonk's workers were toiling through the Fraser Canyon, the CPR launched its own construction assault from the east. The operation slowly worked its way through northern Ontario, then sped across the prairies and up the Bow River valley to the continental divide, which it reached in the fall of 1883. What lay ahead was about 170 miles (270 km) of the most difficult conditions the CPR had to face in the entire project. After a few brief miles of gentle slope, the right of way plunged steeply down the Kicking Horse River valley, dropping about 2,500 feet (750 m) to the banks of the Columbia River at Golden. Timber for this portion of the line was cut at a mill which was part of a large supply camp at Laggan, the present site of Lake Louise. When the right of way reached Palliser, midway between Field and Golden, a steam-powered sawmill was built. Logs were cut in the Kicking Horse valley, skidded to the river with horses and floated down to the mill. There they were cut into timber for

Page 20: The Mountain Creek bridge, which required more than two million board feet of timber. An enormous amount of wood was used during construction of the Canadian Pacific Railway. NAC PA 66610

Page 21: Shuswap Milling Company, Kamloops, c. 1870. BCARS 30627

An unidentified sawmill cutting bridge timbers at the foot of Mount Hermit in Rogers Pass after completion of the Canadian Pacific Railway, 1893. KLM 6-28

The Ottertail Creek bridge on the CPR line east of Golden, 1921. Built with untreated timbers, this bridge and dozens like it were replaced with steel structures within ten to fifteen years. NAC 18905

the five bridges over the river in its 12-mile (19 km) run to Golden. Even more wood was needed to line the tunnels through the enormous clay and mud deposits along the river. For this work, experienced loggers from Ontario and the Atlantic provinces were hired. Many of them elected to stay in BC after construction ended.

At this time there were no mills at Golden, so the Palliser mill provided whatever timber was needed to lay track to Donald, 28 miles (45 km) downstream, and for construction of a 400-foot (120 m) bridge over the Columbia. There construction stopped for the winter, and another sawmill was built at Beavermouth, where the Beaver River flows into the Columbia.

The next phase of the project, through the Selkirk mountains to Revelstoke, was perhaps the most daunting section of the entire 3,000-mile (4800 km) railway. The terrain was steep and rugged, with numerous creeks running through deep canyons into the Beaver River on the eastern slope and the Illicillewaet River on the western slope. Enormous wooden bridges were needed to span these creeks. They would last only ten to fifteen years, but the CPR lacked the funds to build steel structures at the outset. One of the first bridges up the Beaver, at Mountain Creek, was 164 feet (49 m) high and 1,086 feet (353 m) long, and used more than 2 million board feet of timber. The Stoney Creek bridge, the highest railway bridge in the world, was 292 feet (88 m) high, and there were dozens of other bridges.

The Beavermouth mill, a large steam-powered mill, was built and began cutting timbers during the winter of 1884. However, this operation was severely hampered by the second major difficulty presented by the Selkirks: heavy snowfall. James Ross, the engineer for the CPR, visited the site in February and noticed that as the loggers worked, felled trees had to be dug out of the snow before they could be bucked and skidded to the mill. The following winter, not long after the line was completed, it was closed by snowslides. The men in charge decided to construct more than thirty snow sheds, a staggering job requiring 18 million board feet of timber. Most of this wood was also cut by the Beavermouth mill.

During the summer of 1885, railway building crews pushed slowly west through the Selkirks. From the west, Onderdonk, who had completed his government contract to Kamloops, continued under contract with the CPR to build the line east to Eagle Pass. He was within 2 miles (3 km) of his goal by September when he ran out of rails. Six weeks later, on November 7, 1885, the westbound crew reached the same spot, named Craigellachie for the event, just before the first winter snowfall began. The transcontinental railway was built.

Following completion of the railway, most construction mills either ceased to operate or continued to cut auxiliary materials for the railway. It took a few years for mills independent of the CPR to become established, but before long there was a renewed demand for wood. Some mills served the local market and others provided lumber for the CPR and the construction market that was emerging on the prairies. As the railway became operational it consumed enormous amounts of lumber for stations, warehouses, boxcars and other requirements. More ties were needed for sidings and spur lines as they were built at various points along the way. As towns and villages along the route began to grow, they too consumed growing volumes of lumber and other building materials.

After snow slides closed the railway during its first winter of operation, fifty-four snow sheds like this one in Rogers Pass were built at a cost of more than $2 million.
WMCR NG4-255

The first independent mill to emerge from the railway's construction phase was Columbia River Lumber (CRL), a subsidiary of the Canadian Western Lumber Company which also owned Fraser Mills at New Westminster and the Comox Logging and Railway Company on Vancouver Island. The CRL grew out of the CPR's Beavermouth mill. For several years it operated two mills on the Beaver River, then built a new mill on the Columbia at Golden in 1900. This was a strategic location for a large mill to serve the growing prairie market. Vast stands of high-quality timber growing to the south, along the Columbia, could be floated to the Golden mill, ensuring the long-term timber supply that would justify investment in a large modern mill. And with the railway at hand to ship lumber, the mill was in an ideal location to serve the prairie market.

As this market grew, CRL's operations expanded to keep pace. Other mills were built at Kualt and Notch Hill on the CPR near Shuswap Lake, and an extensive railway logging operation was opened at Nicholson, a few miles upriver from Golden, in the mid-1890s. It was the first such operation in the Interior. The first locomotive used at this camp was a small, well-used Baldwin purchased from the CPR, initially named the Countess of Dufferin but known as Betsy in her logging days. The railway operation eventually grew to include three Heisler locomotives, by which time the Countess had been retired, her boiler put to use in the planer mill and her engine in the sawmill, and her wheels used to move lumber around the mill yard. Later these pieces were rescued by a Winnipeg group and the Countess was reassembled and restored for public display.

A horse-drawn sloop with a big load of logs at the Elk Lumber Company operation near Hosmer. It supplied logs to a mill at Elko, which closed in the 1920s. BCARS 24129

The Nicholson logging show was typical of North American railway logging operations of that era. A camp was set up on the west side of the Columbia and a rail line was run north along the benchland above the river, eventually reaching a point across from Donald, about 25 miles (40 km) away. As the rail line was extended farther north, other camps were set up to house loggers. Most of the logging was done in the winter.

Trees were felled with crosscut saws and axes, and skidded to the railway with horses. Sleighs were used in the winter and sloops were occasionally employed in the summer. A sloop consisted of a single set of runners with the tail ends of the logs dragging behind to serve as a brake on downhill runs. Horses for this operation were imported from Alberta ranches. They

Loading logs with a Barnhard loader at Columbia River Lumber's railway logging camp near Golden. BCARS 73492

were replaced regularly. The heavy work of skidding and hauling logs wore them out after four or five seasons of logging, and they were equipped with caulked horseshoes which often injured their legs.

At railside, logs were loaded onto cars with a state-of-the-art Barnhard loader. This steam-powered machine was built in Pennsylvania. It sat on a set of steel tracks laid on the rail cars it was loading, and pulled itself along these tracks with a wire rope. Its engine, boom and winches were mounted on a turntable, the first machine of this type ever built.

The loaded cars were then hauled to the river, where they were dumped and floated to the mill at Golden. The mill too was a state-of-the-art facility, operated by electricity generated in a steam plant fired with planer shavings. It had two double-cutting band mills as head saws, to make the initial cuts in a log. The band saw was a recent innovation, having come onto the North American market in about 1890. It consisted of a looped steel band or belt mounted on two large wheels. To mill lumber with one of these saws, workers mounted the log on a carriage which passed it through the blade. A double-cutting band had teeth on both edges, permitting boards to be cut as the carriage moved forward and back. One of the great virtues of the band saw was that it made a much thinner cut, or kerf, than a circular saw, wasting less wood in the form of sawdust. Head saws were generally used to produce large-dimension boards or cants, which were resawn to smaller dimensions with other narrow-kerf saws.

Resawing was done with a 48-inch (120 cm) gang mill, consisting of a sash or frame (of the type used in old water-powered mills) and as many as twenty or thirty vertical blades which could be adjusted to the thickness required. Logs or cants could be passed through the gang saw, which operated with a reciprocal motion, cutting several boards at once. Gang mills cut boards very accurately and, like band saws, produced narrow kerfs and relatively little sawdust.

The mill was equipped with two edgers, to saw lumber to the desired width, and a planing mill to smooth boards and produce mouldings and other specially shaped materials. There was also a lath mill to utilize the clear, straight-grained wood found in the outer portions of big logs. During this period, lath was used extensively in construction as an interior-wall base over which plaster was laid.

During its peak years, the CRL's mill at Golden shipped 40 million feet of lumber a year. The mill ran until 1927, when much of its timber supply was burned in a fire. Through most of its operating period, other, smaller mills also produced lumber in and around Golden. The Palliser mill continued until 1908, when it ran out of timber. A water-powered mill at Armstrong ran for several years, as did various others. Among them were several at Donald, which utilized timber from along the Blaeberry River.

Timber cut during construction of the railway, and along its route for many years afterward, was obtained from timber berths available in the Railway Belt. This was a 40-mile-wide (64 km) strip along the railway that BC had given to the federal government to help pay for construction. The federal government administered the area and sold timber to logging contractors who worked on the railway, and to lumber companies established later.

In 1886 Thomas Higginson, son of an Ontario Member of Parliament, was appointed BC's Crown Timber Agent and given responsibility for issuing Timber Berths and collecting royalties for timber cut. Higginson earned the distinction of being not only the province's first timber official, but also the first one proven to be corrupt. Eventually his activities became so blatant that in 1897 a public inquiry was called. It revealed, among other things, that Higginson had issued the CPR a "roving" permit to cut timber, any place at any time, in exchange for an unrestricted pass on the railway. The inquiry also implicated several prominent Interior lumbermen who had conspired with Higginson to defraud the government of its revenues. Among them were James McIntosh, a partner in the Shuswap Milling Company at

Sawmill and grist mill at James McIntosh's operaton, the Shuswap Milling Company, built in Kamloops in 1869.
BCARS 9808

Kamloops, a monument to whom still graces the city. John Mara, McIntosh's partner, was a local MP who had helped Higginson obtain his appointment and increases in pay. The greatest beneficiary of Higginson's manipulations appears to have been Joseph Genelle, who built a sawmill at Kualt in 1886, and whose brother Peter later built the first sawmill at Nakusp.

Although the inquiry report was suppressed, both McIntosh and Genelle disappeared from the Kamloops sawmill scene, leaving the thriving community without a supply of lumber. John Shields, an Ashcroft businessman, stepped in to fill the void and set up the Ryans-Shields Sawmill Company in 1899, employing 100 loggers in a dozen camps along the North Thompson. Within a year or two the company became the Kamloops Sawmill Company, and set up sawmills at Enderby and Avola. In 1906 the company was sold again, to C.R. Lamb, a Wenatchee lumberman who was among the first of a flood of US entrepreneurs to enter the province. The operation was renamed Lamb-Watson Lumber. By this time several other smaller companies had appeared in Kamloops, the largest of them being the Thompson River Lumber Company, with a mill on the present site of Pioneer Park.

In the early years of this century, these companies, along with several of their predecessor companies still operating in the Thompson–Shuswap area, became embroiled in a protracted battle over control of river driving and booming rights on the Thompson River. The federal government, in its role as administrator of navigable waters, had granted to a lumber company a charter giving control over a stretch of the Columbia River, a practice common in east-

Peter Genelle's sawmill at Nakusp, 1897. The mill burned in 1906. BCARS 31793

Cedar Valley Lumber's Camp 1 near Fernie, shown here in 1909, was typical of accommodation provided for loggers around the turn of the century. *BCARS 64654*

ern Canada. In Kamloops, the struggle to gain control over the North Thompson River led to the organization of several "river improvement" companies, each made up of various combinations of competing lumber companies. In the end the Lamb-Watson group, now known as Arrow Lakes Lumber, triumphed. It took over control of the river, clearing obstructions, driving pilings and making other modifications to speed the flow of logs, with a booming grounds at Kamloops to sort the logs according to ownership and destination. Lamb-Watson charged what its competitors claimed were excessive tolls for every log that went down the North Thompson.

The completion of the CPR transcontinental line in 1885 stimulated the construction of numerous other branch lines throughout the southern Interior during the 1890s. The CPR participated in this transportation boom mainly in response to the threat of competing lines coming in from the USA. In 1882 the Northern Pacific Railway (NPR) had completed a transcontinental line through northern Idaho and Washington, as did the Great Northern Railway in 1893. With the completion of the CPR, access to the mineral-rich Kootenays was greatly improved, although neither the CPR or the NPR actually penetrated the area. Both lines cut across the natural grain of the country, including its navigable waterways. What evolved quickly was a transportation network in which east–west traffic went by rail; north–south operations were handled by steamships on rivers and lakes. To fill in the gaps in this network, railway companies began building branch lines into southern BC to tap the markets created by the growth of the mining industry throughout the area. Construction of these lines, in turn, stimulated development of forest industries in the same way it had during the building of the CPR main line.

In the west Kootenays, for instance, the first sawmill of any significance was the one Peter Genelle built at Nakusp in 1891. The Genelle family was typical of many who founded Interior lumber companies in the wake of CPR construction. Three Genelle brothers—Joseph, Jack and Peter—along with their two sisters, Addie and Sadie, were the five oldest in a family of fourteen children from Thessalon, Ontario. When their father died, they went west to hack ties for the CPR. Working as a tight family unit, with the men running crews of tie cutters, the women operating camp kitchens and the children doing chores, they quickly amassed a sizable bankroll. Upon completion of the railway, they invested their earnings in the sawmill at Kualt, and from its earnings a second mill was built at Yale. Peter and Jack then decamped for Nakusp, leaving their older brother Joe to his misadventures with Thomas Higginson, the corrupt timber agent.

Genelle's Nakusp mill operated at a capacity of 10,000 board feet a day, then quickly tripled in size, in part to supply materials for the Nakusp and Slocan Railway. A second mill was built at the bottom end of Arrow Lake in 1892, to supply timber for construction of the Columbia and Kootenay Railway. By the end of the decade, this company, now known as Yale-Columbia Lumber and still owned by the Genelle family, owned mills at Nakusp, Robson, Rossland, Cascade, Rock Creek and Deadwood. Five years later, in 1905, the Genelles were bought out by Bowman Lumber of Minneapolis, Minnesota, which already owned extensive timber berths around Revelstoke. Later that year, Bowman amalgamated with Elk Lumber, another large sawmilling company in Fernie, making it the largest operation in the Interior with a combined daily capacity of 1.5 million board feet. The industry was barely underway, yet corporate concentration had already become a fact of life.

During the same period, several large mills were built in the timber-rich Slocan Valley, including Hill Brothers and the New Denver Lumber mills on Slocan Lake, and the Wineau mill 13 miles (21 km) down the Slocan River. At Kaslo, on Kootenay Lake, the Buchanan

Following pages: The Baillie-Grohman sawmill at Canal Flats, 1887. *UBC/BG*

Lumber Company was started by G.O. Buchanan, a prominent Interior conservationist, in 1890. All these mills served booming local markets created by mine development, steamship construction and the growth of new communities throughout the Kootenays.

In the east Kootenays at what is now Canal Flats, an unorthodox venture established the first steam-powered sawmill in that region. William Adolph Baillie-Grohman was an English author and aristocrat, cousin to the Duke of Wellington and blessed with an entrepreneurial spirit, who had hunted in the Kootenay River valley where it flows out of the Rocky Mountains. He noted the broad plains at Kootenay Flats where the river emerges from the mountains, less than a mile from Columbia Lake, the source of the Columbia River. He also noted that this 50,000-acre (20,000 ha) expanse of potentially productive farmland flooded every year, rendering it useless for agriculture.

Baillie-Grohman conceived the idea of diverting the Kootenay River into Columbia Lake, which is only 11 feet (3.3 m) below the level of the Kootenay, and letting it drain through the Columbia River. This would leave the flats dry and useful. He persuaded the provincial government to go along with this scheme and returned to England to raise the money for a diversion canal. But the CPR objected, fearing increased flows in the Columbia would

Left: Mill crew with the head saw at Brydges & Fisher Lumber, Nelson, 1899.
KMA S9.9.65 274

Below left: Brydges & Fisher Lumber's planer mill.
KMA 59.7.9 270

interfere with the rail line below Golden. The federal government intervened and refused Baillie-Grohman permission to divert the Kootenay, but permitted him to connect the watersheds with a canal and a lock.

As he admitted later, he should have pulled the plug at that point. But entranced with the engineering challenge, he forged on. In 1886 he arrived at Golden on one of the CPR's first commercial runs, equipped with a modern steam-powered sawmill to cut the timber he needed to build the canal and its lock. His first misadventure occurred when he moved his equipment up the Columbia from Golden. The only transportation available was a vessel of dubious buoyancy, the *Clive*, still under construction. It consisted of an old barge used during construction of the CPR, the boiler from a Manitoba steam plough and an assortment of other parts from various worn-out tugboats. The owner of the *Clive* did not have the boat-building skills to assemble his craft, so Baillie-Grohman and his sawmill mechanic first had to finish the job. That done, they headed off up the Columbia, taking twenty-three days to complete the 100-mile (160 km) journey. The *Clive* boasted no cabin and rain fell steadily throughout the voyage, so Baillie-Grohman spent much of the trip sleeping in the firebox of his sawmill's boiler.

Undaunted, he arrived at Canal Flats, set up his sawmill, dug a canal and installed the lock. But the flats continued to flood; his land reclamation plan was a failure. He eventually returned to England, leaving behind a canal which slowly filled in.

While Baillie-Grohman did not realize his dream, he did bring some useful qualities to the infant forest industry—resourcefulness and unflagging determination even in the face of terrible obstacles, qualities that have been required by successful loggers and lumbermen ever since. For example, during the course of his project he needed a tugboat of his own. He obtained a small steam launch, the *Midge*, which had been built of teak in China and was headed for Norway, by buying it off a ship on the Atlantic—avoiding duty by declaring the vessel an agricultural implement. He then shipped her across the continent on the Northern Pacific Railway to a point 39 miles (62 km) from the Kootenay River. From there, a large crew of white and Native workers carried and dragged her overland to Bonners Ferry on the Kootenay, from where she sailed up to Canal Flats. The Natives in the area were fascinated by the *Midge*, especially her steam whistle. Baillie-Grohman persuaded them to provide him with cordwood for her boilers in exchange for towing their canoes up the river, and for allowing them the thrill of blowing her whistle.

Further development of more conventional transportation facilities extended the forest industry into the southern end of the Columbia–Kootenay valley. In 1898 the CPR-owned British Columbia Southern Railway was built from Lethbridge through the Crowsnest Pass to Fernie, and on to Cranbrook, Kimberley and Creston at the south end of Kootenay Lake. The following year the CPR built a large sawmill at Fernie to supply lumber to its operations in southern BC and Alberta. The Crow's Nest Pass Coal Company, a subsidiary of Great Northern Railway, built several mills east of Fernie to provide lumber and ties for its mining and railway operations.

In Cranbrook, a small sawmill had operated since the 1870s. In anticipation of the boom expected to follow railway construction, two more mills were built in 1898—Cranbrook Lumber Company and Cranbrook Sash and Door. The latter would survive under the name Crestbrook for more than a century.

Similar developments followed the extension of rail lines into other areas. The Columbia and Western Railway was laid from Trail north to Castlegar, and southwest to Midway, with plans of forging on to Hope. Small mills built to serve local markets followed in its wake. The same thing happened in the north Okanagan, with construction of the Shuswap and Okanagan Railway in 1892. Most of the lumber and heavy timbers used in building the railway were provided by a mill built in 1883 by Alfred Postill on a ranch he owned near Winfield. It was the first sawmill in the area. Two years after the railway arrived, S.C. Smith, a Vernon lumberman, moved his mill from Okanagan Landing to Enderby. The mill burned and was rebuilt at the turn of the century, after which the operation employed 1,000 men in the mill and at camps up the Shuswap River to Mabel Lake. Farther south, the Kelowna Sawmill, built by Bernard Lequime and David Lloyd-Jones, began operating in 1894. This mill would eventually form one of the foundation stones of Kelowna's Simpson complex.

Meanwhile, in the northwestern corner of the province, the gold rush was playing itself out in a final extravaganza—the Klondike stampede. During the fall and winter of 1897–98 more than 100,000 novice miners flooded into the area. Several routes to the Klondike passed through northern BC, and at various points along the way boats and barges were needed to move people and equipment down rivers and across lakes. Sawmills were built at Atlin and Bennett, the latter owned by the Victoria-Yukon Trading Company, which also operated logging camps. By the fall of 1899 the rush was over. An estimated $50 million in gold had been recovered, about the same as it had cost the gold seekers to reach the Klondike.

By the turn of the century the Interior forest industry, with the exception of Columbia River Lumber at Golden, had evolved only to the point of serving the rapidly growing local markets. Most of these mills and woods operations were owned by resident entrepreneurs who, lacking the capital to develop large-scale ventures, had to generate revenues from their existing operations to finance gradual expansion. The CRL, through its parent company, was affiliated with the most powerful and influential lumbermen in BC, if not Canada. It had

Horse-drawn sleds haul logs to the Atlin sawmill, 1899.
BCARS 61882

The Atlin sawmill, 1899. The mill provided Klondike miners with lumber for boats, buildings and sluice boxes.
BCARS 61890

ready access to investment funds and the latest, most efficient technology. The Genelle brothers, on the other hand, probably the most successful resident entrepreneurs of this period, were more dependent on their own earnings and the resources of local investors.

A prime example of what this meant in practice was CRL's purchase of a double-cutting band saw for its mill. In 1900 this was a radically new method of cutting lumber, far more efficient than the circular or gang saw. While the idea of the band saw had been around for decades, it was not until the late 1890s that functional band blades were available. Before the development of steel alloys, band saw blades strong enough to use in head saws could not be produced. When they did come on the market just before the turn of the century, they became very popular. By 1904, Simonds, one of the largest saw manufacturers, was selling band saw blades in nineteen different tooth configurations.

Skidding logs with sloops at Mabel Lake, 1898. EDMS 462

One of the prime advantages of a band saw is that it wastes less wood in sawdust than a circular saw, particularly when milling large-diameter logs. In order to cut a big log, a circular saw must be as large in diameter as the log, and it must be thick enough to bear the stresses of cutting through such a large amount of wood. Consequently, the kerf of a circular saw is several times that of a band saw that does the same work. This advantage was significant, even in the early stages of Interior sawmill development. When timber harvesting began in the Interior, there were substantial stands of large-diameter trees at various locations. Throughout the Interior wet belt, Douglas fir up to 8 or 9 feet (2.5–3 m) in diameter were common, and big cedar and hemlock were also found in these stands. Farther south, in the ponderosa pine forests, trees 5 to 6 feet (1.5–2 m) across were harvested. North along the Fraser River, 4- and 5-foot (1.2–1.5 m) spruce were widely available. Interior timber was often as big and valuable as coastal timber, although there was not as much of it. The market for all of this wood was strong, so converting prime Douglas fir or white pine logs into sawdust was not nearly as profitable as cutting them into salable lumber. Owners of the big interior mills were interested in state-of-the-art machinery such as band saws right from the early days.

◄

Fallers stand on springboards after completing an undercut in a big cedar at Adams River Lumber. The next step was to use the crosscut saw to make a falling cut on the other side of the tree. BCARS 73479

The situation was somewhat different when it came to logging equipment. Throughout most of the world, logging was still done almost exclusively with human and animal muscle power. In some areas, such as the cypress swamps of the southern USA and the coastal rain forests of the Pacific Northwest, mechanized, steam-powered technology was developing quickly to handle the big logs that had to be moved over rough or soft ground. This was not the case in most of the Interior. Smaller concentrations of big timber made the highly mechanized operations which were evolving on the Coast uneconomical in the Interior. Here, the most efficient means of skidding logs short distances was with horses. Most of the available usable timber grew within horse-skidding distance of a river or lake. In the rare instances where this was not so, such as in CRL's timber stands on the west banks of the Columbia, a small railway might be needed, usually to move logs to the nearest river rather than directly to the mill.

At the dawn of the twentieth century, all these factors influencing the birth of the Interior forest industry were about to change dramatically. The market for forest products was about to explode. The procedures for obtaining timber from the BC government were about to be altered profoundly. And a technological revolution which would still be in force a century later was already beginning. On top of all that, the social and economic development of the province was about to swing into high gear. The Interior forest industry, as both a beneficiary and agent of that development, was about to take off.

CHAPTER 3

Growing Pains

The Early 1900s

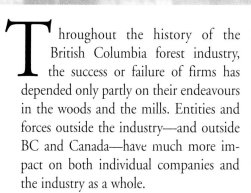

Throughout the history of the British Columbia forest industry, the success or failure of firms has depended only partly on their endeavours in the woods and the mills. Entities and forces outside the industry—and outside BC and Canada—have much more impact on both individual companies and the industry as a whole.

In the early years of this century, a series of events occurred in the United States which had a dramatic effect upon the BC forest industry, particularly the Interior industry. In the 1880s a radical new idea about forests emerged in the eastern US and quickly spread across the continent—the idea of the permanent forest.

Until this time, conventional North American wisdom held that the great untouched continental forest was a resource to be harvested on a one-time basis, after which the arable land would be used for farming and ranching, and the remainder left to its own devices. Timber was looked upon in essentially the same manner as gold or coal: something to be extracted. After it was gone, the industrial apparatus employed in its utilization would be relocated or disposed of. The question of what would occur when all the timber was harvested was not answered and seldom asked.

The new idea was that forests could be renewed and maintained in perpetuity, and that the industries engaged in harvesting and processing timber could become permanent features in local and national economies. People who adopted and professed these ideas—often with evangelical fervour—called themselves conservationists. By the 1890s they had become a powerful force in US and Canadian politics, and their work had culminated in the creation of national forest reserves in both countries. The purpose of these reserves was not to remove them from commercial exploitation, but to ensure they were harvested on a schedule and as part of a "scientific" management system which ensured a continuing supply of wood. In the USA most of these National Forests were located in the western states; in Canada they were established by the federal government in areas under its jurisdiction, primarily those portions of the Northwest Territories that would become Alberta, Saskatchewan and Manitoba.

The effect of these actions was to refocus the energies of

the established North American lumber industry. It had developed in eastern Canada and the northeastern USA, worked its way through the forests of those regions and headed west in the wake of railway development and settlement. Speculators went ahead of the industry, obtaining manageable blocks of timber from the railway companies, which had obtained land grants to aid in construction of their lines. Cut off from vast areas of western US forest by the designation of National Forest land, the speculators looked to BC and found a provincial government desperate for revenue to pay for programs it hoped would get it re-elected to office.

Before this time there had been two principal ways to obtain timber in BC. It could be purchased from railway companies, along with the land on which it grew, or it could be obtained from the provincial and federal governments in the form of leases, licences or berths allowing the harvesting of a specific volume or stand of timber. Outright purchase was a more costly method but its end result was a salable commodity, whereas the other method involved various harvesting tenures that could not be sold. In 1905 the government of Richard McBride dealt with the problem by bringing in a new form of Special Timber Licence, which could be sold. Overnight a small army of timber stakers spread throughout the province, securing leases on most of the merchantable timber (by the standards of the day) in the province. This in turn brought many established lumber producers, along with their financial resources, into BC from both eastern Canada and the USA.

Coincidental with this development was an active settlement program on the Canadian prairies. A joint promotion by the CPR and the federal government attracted tens of thousands of would-be farmers, and their presence justified the construction of an extensive railway network throughout the west. Hundreds of small towns were built, and as the farmers became established, thousands upon thousands of wooden farm buildings—houses, barns, sheds and so on—were constructed. In those parts of BC with stands of suitable timber and connections to the rail net—essentially the Kootenays and parts of the Thompson–Okanagan region—numerous large sawmills appeared in the first decade of this century. Many of them were also able to supply a booming local market based on mining and agricultural development. The major market, however, was the prairie rail market, an external event to which sawmill operators, many of them new to BC, responded.

Pages 38–39: A Mabel Lake logging crew in their camp, c. 1900. EDMS 2615

> A mill in those days, to get out that much lumber, had to have a big crew in the bush, a lot of men and a lot of horses. To fall the trees it took two men and the old crosscut saw, then somebody came along with an axe to take off the branches. Then a skidder and skid team came to skid them onto a landing. In those days it was all man power and horse power. With horses and the old crosscut saw it would probably take from seventy-five to a hundred men out in the woods to keep a mill that size [70,000 board feet a shift] running. There would probably be about sixty horses out in the woods with them, and every time you put a team of horses out there you had to put a man with them, to drive them.
>
> In lumber camps of this size the foreman's office also served as a store where the men could buy items they needed. The foreman and the timekeeper also acted as store clerks. Bush foremen were usually excellent in the woods and good with the men, but many could neither read nor write.
>
> Once our friend Harold Beattie was acting as clerk, but he had to go to town. So Mackie, the foreman, said he would look after the commissary for him. When Harold got back he found that Mackie had had the men sign their names and he had drawn pictures of the purchases beside the names. One fellow bought a pair of drawers, a pair of pants and two boxes of snuff. Mackie had drawn two pairs of pants, one of which had a cross on for drawers and the other with no cross, just pants. He explained to Harold that the cross on one pair meant drawers and two circles were the two boxes of snuff.
>
> —Jack Aye, who worked in Cranbrook area sawmills and logging camps in the early 1900s.

The Lamb companies and Columbia River Lumber (see Chapter 2) were leaders in this development, and they were soon joined by others. Some of them established new mills or took over existing mills in the Okanagan and expanded them. Typical were the three Smith brothers from Renfrew, Ontario, who travelled west on the CPR in 1901. At Armstrong they bought a small mill from the Wood & Cargill Company for $3,000, and renamed it Armstrong Sawmill Ltd. When they took over the mill it consisted of a circular saw and an edger. They purchased horses, sleighs and wagons to set up a logging operation and launched a business which survives, in much altered form, almost a century later.

> Adolph's mill was a circular-saw mill. They had two circular saws, one on top of the other, for big logs. That was their head rig and carriage. I guess there would be another seventy-five to a hundred men working in each of those mills, when they were operating. There was the crew in the mill itself and then the men on the platforms handling the lumber at both the planer and the mill. Then there was the crew in the yard piling the lumber, as well as the men in the planer mill.
>
> In a mill cutting 100,000 board feet there would be three or four men, four probably, that would pull the lumber off the green chain. In most of these mills it was East Indians that did that work. There would be one boss that would take a contract to look after the green chain and it was all done through him. He hired whatever help he wanted, and if he needed four men or if he could get by with three, well, that's what he did.
>
> In the mill yard the lumber was dry-piled in rows and the alleys were all numbered. It had to stay out there for a year with the air circulating through so it would dry, and they would build a board roof over it. Then the next year, or whenever they wanted to use it, they had to send men out there again to put the pile on a roll-off wagon and take it to the planer mill.
>
> It was put through the planer and was either put directly into a boxcar on the spur line next to the planer, or loaded on a buggy and taken to the dry shed. The buggies were three-wheeled and were pulled around the platform by one horse. A whistle was blown ten minutes before quitting time to give the teamsters and skinners time to put the horses in the barn. Some of the horses pulling these carts got so smart that when the whistle blew, they would take off for the barn on their own and it was too bad for the buggy and the lumber; a smash-up took place.
> —Jack Aye

At the other end of the Okanagan in 1905, Hugh Leir, a young British immigrant, jumped into the sawmill business with a small steam-powered mill he set up on Ellis Creek. Leir had spent a couple of years at Keremeos, where he built and operated a ferry on the Similkameen River. He moved to Penticton when he learned of the Southern Okanagan Land Company's plans to embark on an orchard business. He got a contract from the company to cut 3 million feet of lumber for irrigation flumes, scrounged parts from numerous locations as far away as Vancouver, and put together a mill.

By 1910 he had cut his way through the limited timber supply of the area and relocated his mill at the south end of Okanagan Lake, a short distance down the Okanagan River. He accurately predicted the construction of a rail line past this location a few years later. His new mill was a much improved facility, capable of cutting 30,000 feet a day, and was incorporated as Penticton Sawmills Ltd. The initial timber supply for the mill stood at the head of Four Mile Creek, more than 3 miles (5 km) from the lake. Leir built a chute using long poles down which logs were shot to the lake. The chute operation soon became a local landmark as the logs, travelling at 60 miles (100 km) an hour, left a plume of smoke visible for miles. Where the chute crossed the Naramata road, a watchman armed with a bugle warned travellers of approaching logs. The logs burst out of the flume into booms on the lake, then a tug towed the booms to the mill. Aware of the limited timber supply in the dry south Okanagan, Leir spent the next few years patiently acquiring most of the timber rights within a 100-mile (160 km) radius of Penticton, ensuring he would never face serious competition in the local market. His timing was good, too: Penticton underwent a major development boom just after the new mill opened, and Leir was able to supply most of the lumber required by the new community.

In 1906, at about the same time Leir was launching his business, the Interior's first second-generation lumberman entered the business on Kamloops Lake. John Shields's son James built the Monarch Lumber mill at Savona with his father's financial backing. Shields had learned the business managing the family's Shuswap Shingle and Lumber mill near Sicamous, and opening up the Shieldses' Sovereign Lumber Company in the Cariboo.

Farther east, up the South Thompson watershed at the small farming community on Shuswap Prairie, just below the outlet of Little Shuswap Lake, J.P. McGoldrick, a well-established Spokane lumberman, obtained a lease for a mill site from the Chase family. On it he built what was by far the largest mill in the Interior at that time, the 160,000-foot-per-shift Adams River Lumber Company. It was the first major industrial operation in the BC Interior.

The mill, which began cutting in late 1909 or early 1910, was powered by a 1,000-horsepower steam plant fired with shavings and sawdust. It was a state-of-the-art mill complex with double-cut band mills and the requisite edgers, trim saws and planers for turning out a full range of lumber products. A town was built to house the three hundred millworkers, and the mill and town were supplied with electricity generated by the steam plant. A mile-long rail spur was built off the CPR main line to load lumber destined for the prairie market.

The mill's timber supply was located above and along the Adams River. McGoldrick initially obtained 43 square miles (112 km²) of timber on the Adams River and another 36 square miles (94 km²) on the plateau above the watershed. In 1912 he bought another 25,000 acres (10,000 ha) of timberland.

In an unusual move, logging commenced at the back end of the timber holdings, at Tum Tum Lake, about 75 miles (120 km) from the mill. Before it could begin, an extensive supply and log-transportation system had to be put in place. A tote road was built from Shuswap Lake, which was navigable from Chase, for 7 miles (11 km) up the lower Adams River to Adams Lake, where a supply depot and wharf were built. Another depot and some bunkhouses and a boarding house were built at the head of the lake, 40 miles (64 km) away,

Adams River Lumber's sternwheeler Helen *towing logs on Adams Lake.* BCARS 73484

An Adams River Lumber logging camp on the upper Adams River, c. 1910. BCARS 73473

and a 30-mile (48 km) tote road was built along the upper Adams into the timber licences. A small steam-powered mill was set up near the upper logging camps to supply lumber to the logging operations. A 200-ton sternwheeler, the *Helen*, was built on Adams Lake to tow logs and transport men and equipment, and a smaller 100-ton tug, the *Crombie*, was built to perform similar functions on Shuswap Lake.

Most of the logging was done in the winter, from a series of camps. Logs were skidded to the lakes or river with horses, sometimes with the use of chutes, and were then driven down the upper Adams to the lake. They were towed to the lower Adams, and a splash dam at the foot of Adams Lake, built by the company at the start of its operations, flushed them down the lower Adams. Crews worked a splash dam by storing up water, then letting it go in one powerful burst to push logs along. The dam prevented the migration of fish into the upper Adams watershed, nearly wiping out the spectacular sockeye salmon run above the lake. Use of the dam also severely damaged the salmon spawning grounds on the lower Adams. In 1914 the problem was compounded when railway construction triggered a rock slide at Hells Gate in the Fraser Canyon, preventing the passage of any salmon past that point. In 1947 the Adams Lake dam was dynamited, and around the same time permanent fish ladders were built at Hells Gate. These were the first stages of a decades-long project to restore the prolific Adams River sockeye runs—one of the most significant resource rehabilitation projects ever undertaken in BC.

The Adams River logging show was a conventional high-grading operation of its time, with only the commercially usable fir and cedar logged and at least half of the timber left behind. Within three or four years most of the timber within skidding distance of the lake was depleted and the company turned to other methods. On Bear Creek a flume was built to float logs 11 miles (17.5 km) to the lower Adams—probably the longest flume of its kind in the world at that time.

Flume construction was a highly sophisticated and expensive proposition—it cost about $12,000 a mile. At Bear Creek, trestles up to 80 feet (24 m) high were built to hold the evenly descending, V-shaped flume. It was 15 feet (5 m) across at the top and built with two

overlapping layers of inch-thick lumber. Logs were fed into the flume at various loading sites along its route, and water was added at several points to maintain sufficient levels to float the logs. Walkways were built along some portions to facilitate maintenance and log movement.

While costly to build, flumes were a very efficient means of moving logs. The Bear Creek flume transported them at a speed of almost 45 miles an hour (72 kph) with a crew of only five or six men, who could send down as much as 10 million feet of logs a month. The company built and used several other, shorter flumes during its operation.

One of the keys to a successful flume setup was the speed with which logs could be removed from the end of the flume. At that time most communications systems were primitive or nonexistent, and if a jam occurred at the outlet, a lot of logs could be sent down the flume before workers could be notified to stop feeding them in. The Bear Creek flume was highly successful in this regard: logs coming out of it were thrust into the lower Adams River, swept downstream by fast-flowing waters, collected in a bag boom in Shuswap Lake and towed to the mill.

◄ *The Bear Creek flume built by Adams River Lumber, c. 1910. A view of the flume showing a walkway which provided access for maintenance. BCARS 73475*

In 1903, a few years before the Chase mill started up, another US-owned mill was opened, farther east along the CPR main line at Revelstoke. The Bowman Lumber Company of Minneapolis, the major distribution centre for US lumber sales, took over three small mills in and around Revelstoke. These mills formed the basis of Revelstoke Building Materials, a company that eventually acquired lumberyards, concrete plants and other operations throughout western Canada, and which still operates today.

> The Adolphs had a lath mill, too, for part of the time. This used to make laths about two inches wide and about three-eighths of an inch thick. They were used in construction to nail on studding and then be covered with plaster. They had the planer, too, and practically all the different kinds of knives to make moulding, siding, everything required in building.
>
> When they could get the clear pine in those big logs they could make boards two feet wide, clear, not a knot in them, that's where the money was. It was used for furniture, shelving, all kinds of finished products. Now people buy knotty pine, but then there was no value to it. You could use pine for most things, for siding or flooring. It made a nice floor, if you didn't let anybody in there with hobnail boots on. In those days, though, you never used it for studding or joists because it was so soft that it didn't hold nails well. Nowadays they are using a lot of it for studding, they are even using jack pine for that. In those days the people in the markets they had locally and in the USA and eastern Canada didn't want that at all.
> —Jack Aye

◄ *J.P. McGoldrick's Adams River Lumber mill at Chase. The mill was by far the largest and best managed industrial operation in the Interior in 1919. KM*

Because of its magnificent stands of timber, the Interior wet belt south of Revelstoke attracted a number of early ventures in the early twentieth century. At the head of Trout Lake, the source of the Lardeau River, an English company built the large, ill-fated Canadian Timber and Saw Mills Company, with a 60,000-board-foot-per-day capacity sawmill, in 1904. The intention was to cut cedar and ship it down the lake by barge to Gerrard, where it would be reloaded onto the newly built Kootenay and Arrowhead Railway. From there it would be hauled to the head of Kootenay Lake and loaded onto barges which connected with the rail network farther south. But the owners seriously misjudged the situation. The mill was too remote from its markets, and its lumber required too much reloading to make the operation profitable. Within a short time the company was bankrupt. The local sheriff was in the process of auctioning off its assets when two directors stepped in, paid off its debts and rebuilt the mill at Gerrard, on the railway, renaming it Canadian Pacific Timber Company. The new operation was marginally more successful, but survived only until about the end of World War I when it too succumbed to poor planning and even worse management.

A few other sawmills of historical importance were built in the Kootenays at this time. In 1904 Creston Sawmills started. The same year, to finance expansion of his sash and door factory, Tom Leask went into partnership with Jack Slater, an Ontario lumberman. The operation ran through various partners, then reappeared in 1904 as Cranbrook Sash and Door. Its principals included various members of the Slater family and Allan Nicholson, who would later serve as Timber Controller during part of World War II.

> An English company had a mill at Gerrard and they had no one out there to supervise the work, so they were just robbed blind. The chaps who were running it had a padded payroll. They shipped all the best timber somewhere without the company knowing it. If it had been properly looked after, it would be there today, I'm sure.
>
> The mill was running in the First World War. They had a lot of East Indian Sikhs. They were great men for being around the mills. That was in the early days. They were all reservists in the Indian army and they all left in '14 to go back to it.
>
> About the end of the First World War, it was the most up-to-date mill in BC and they had one of the first band saws.
> —*Maitland Harrison, who moved to Howser after World War I and ran a logging camp along the lower Lardeau River for many years.*

A similar process occurred at Edgewood, on Lower Arrow Lake. In 1896 William Waldie, whose family operated a sawmill on Georgian Bay in Ontario, arrived in the Kootenays to seek his fortune mining. He did well and bought a partnership position in a new sawmill venture, the Edgewood Lumber Company. Because of his Ontario experience, his partners insisted he become manager of the new operation. Like the owners of the Trout Lake mill, the Edgewood group miscalculated and set up a sawmill which was not served by a railway line, which meant that lumber had to be barged to a railway. In 1909, after a rail bridge was built across the Columbia River just below the outlet of Lower Arrow Lake, the sawmill was moved to Castlegar and enlarged.

The new setup was typical of the larger mills being built at that time. It was constructed on two levels, with boilers and steam engines on the first floor and head saws, edgers, trim saws and planers on the second level. It also had a lath mill, a shingle mill, electric generators and dry kilns—probably the first ones in the Interior. Their primary purpose was to reduce the moisture content in lumber and thereby lower its weight, which would in turn lower the cost of shipping it by rail. The mill cut all seven species found in the wet belt forests of the Kootenays: fir, cedar, larch, hemlock, cottonwood and white and ponderosa pine. A wide range of products were produced, including timbers, lumber, lath and mouldings in scores of patterns. The operating principle of this type of mill was to cut the most valuable products from each and every log. With a hundred millworkers, it produced 65,000 feet a day and its products were sold throughout Canada and the USA.

Logs were towed down Lower Arrow Lake and a short stretch of the Columbia by the *Elco*, a tug built at Castlegar, and into the

The tramway on which logs were moved from the Shuswap River into the Rogers Lumber sawmill at Enderby. EDMS 2747

Lambert Lumber's sorting bins near Taghum, 1911. KMA 79.2

boom pond where a jack ladder hauled them up a ramp to the mill floor. Another hundred employees were engaged in logging, providing about half the mill's needs; the other half was purchased from independent loggers and farmers. Logging operations consisted of horses skidding logs to chutes running into the lake. Eventually, when logging progressed to Whatshan Lake, a 5-mile-long (8 km) flume was built to move logs down to Arrow Lake. By 1928, having bought out his partners, Waldie had changed the name of the company to William Waldie & Sons, which continued to do business for another twenty-five years before it was sold to Celgar.

John Bell was a mill builder and a sawmill man. He and Dad formed A.G. Lambert and Company and they started in at Taghum and up Sproule Creek. The wholesale lumber yard and the planing plant were along the railroad at Taghum, and the logging and sawmill operation was up Sproule Creek. They operated from 1905 'til 1933 in that one area.

There were seven sawmills built up Sproule Creek. The first one was a mile and a half up, and the last one in the main fork was eight miles up. That's the one I worked in. And when it was all cut out in the main fork, then they had two different mill sites in the north fork, so that's how he got started in the lumber production end of the business. That was in 1905.

All the lumber came to that planing plant site from the Sproule Creek sawmills by flume. It came down over the hill and around the corner to the slide-off table and that was the water from Sproule Creek. They flumed as high as four thousand feet in a nine-hour shift from the mill eight miles up. It took two or three different sorting gaps manned by two Chinamen. Friction in the bottom of the flume holds the water back as it flows, and the top of the water falls ahead so it makes waves down the flume. When the water gets a little low on the flume the waves don't float the lumber. So it had to be fed half a dozen times in that eight miles down the creek, because lots of it got thrown out by the lumber itself. The flume had fourteen-inch sides, sloped at forty-five degrees, twelve-inch bottom, and when running full of water will float two-by-twelves, three-by-twelves, square timbers and so on.

In a nine-hour period we'd have taken the whole cut of the sawmill plus ten or fifteen thousand feet of lumber from the piles that were alongside of the flume cut the fall before. The fluming had to be done in the spring at high water. We sawed from the first of April to the first of July because there wasn't enough water at the headworks up at the top end to flume lumber after the first of July.

—*George Lambert worked in the Kootenay lumber business all his life. He was born in Rossland in 1902 and he died in 1991.*

Following pages: The Lambert Lumber Company's mill yard at Sproule Creek near Nelson. The lumber flume ran down the hill behind the rail line, which was known locally as the Push, Pull & Jerk Railway. KMA 85.376.3 280

Nearby, in the Slocan Valley, a number of other large mills commenced operation. W.C.E. Koch acquired extensive timber rights on the Slocan and Little Slocan rivers and in 1904 built a mill cutting 40,000 feet a day at Koch, on the Slocan. In the south end of the valley, at Crescent Valley, one of the largest mills in the Kootenays was built by Patrick Lumber, owned by a family which later became famous throughout the west for their hockey prowess. Patrick bought timber on the Little Slocan and, during the first winter's logging in 1907–08, cut 13 million feet of logs and drove them down the river to the mill site. Farther north, in the same year, the Silverton Lumber and Power Company started a mill. Halfway between the upper Slocan and Nakusp, at Summit, a Fernie sawmill owner, J.R. Boynton, started a tie and lumber mill, along with a pole business.

The development of telephone, telegraph and electrical systems in North America around the turn of the century gave rise to another forest industry which flourished in the BC Interior, pole cutting. Typically, pole yards were set up at strategic locations to which poles could be brought for sorting, grading, seasoning and shipping by rail. Orders were usually placed for substantial numbers of poles of a particular size so it was not common to ship them directly from the woods to the customer.

For the first few decades of the business, poles were "made"—cut to length and peeled—in the woods. Cedar was preferred, although pine and fir were used occasionally. Pole makers generally worked alone, each with his own strip of timber. In mixed stands the poles were usually felled and skidded out first to avoid breakage. Trees had to be felled carefully, then neatly limbed and peeled. A pole had to be straight and perfectly sound, with any butt rot bucked off, and minimum top diameters were specified. A skilled pole maker could finish a 50-foot (15 m) pole in an hour or less.

One of the first pole yards established in BC was an operation at Nakusp, owned by the Lindsley Brothers Pole Company of Spokane. It was set up and managed by Alec Yoder in 1905. Poles were collected from the surrounding area, the best stands being located in the Monashee Mountains, where two of the largest companies—both of them still operating—established yards. Carney Pole was founded in Iowa in 1905 and set up its first Canadian operation at Chase in 1908. Two years later it opened yards at Enderby and Grindrod. Finished poles were driven down the Shuswap River to the Enderby yard. In the 1920s this yard was shipping 400 rail cars, each containing 75 to 100 poles, every year. A second company, Bell Pole, was founded in the USA by M.J. Bell in 1909. Like Carney, it soon established operations throughout the southeastern Interior, and still maintains several pole yards.

Within a few decades, as logging trucks and peeling machines came into use, poles were no longer being made in the woods; instead they were finished at the pole yards.

One of the last southern Interior rail lines to be built, the Kettle Valley Railway running from Hope to Midway, was started in 1906 and precipitated the construction of a series of sawmills along its route. One of the earliest was the Nicola Valley Lumber Company, established at Canford, west of Merritt, in 1906. Its founder was Andrew McGoran, an Ontario

Skidding poles along the Shuswap River. They were peeled and cut to length in the woods and then skidded to the river for the spring log drive to a pole yard in Enderby.
EDMS 2742

lumberman who staked more than 100 square miles (260 km²) of Special Timber Licences during the staking frenzy then taking place. McGoran had partners from Kaslo who supplied the mill with lumber and timbers from their Kootenay mills. The Kettle Valley Railway was Nicola Valley Lumber's first big customer. At the peak of construction, which went on for more than ten years, the mill was shipping ten carloads of finished lumber to the railway every day. Then, at the height of this activity, the Canford mill burned. McGoran moved to Merritt to start over. He sold the remains of the Canford mill to Henry Meeker, a recent immigrant from Connecticut, who rebuilt the operation and named it Nicola Valley Pine Mills. The business later grew to become one of the largest sawmills in the Interior.

The Hanson Lumber & Timber Company pole loading site west of Hazelton, c. 1925. The poles were shipped after they had been driven down the Skeena River to this site. BVM P271

> The W.W. Powell match block mill had a logging camp out on Lost Creek, below Salmo. It was on the old Hanna limits and they summer logged there with trail chutes. They were made out of fir or tamarack or hemlock, and they put down the ties and they hewed one side of this timber with a broadaxe. They were probably thirty feet long and they put them in so that they made a cradle, with one side hewed smooth. They would skid to this chute, and then they would roll them into the chute. They would have a single horse there with a draft chain and a swamp hook. They put the hook in the back log. That would be a big log so it wouldn't be pulled out of the chute sideways. They would have a space between the logs so that one bumped the other one, and bump bump bump, away they went. You'd haul that out to where it was steep enough it would run by itself. It would run down to the next flat or semi-flat and another horse would pick it up and take it on the next way and so on and so on.
>
> —Russell Fletcher, who was a logger in the west Kootenays. He moved to Nelson when he was ten, in 1908, and began working as a logger after serving in World War I.

The announcement of another railway, the Grand Trunk Pacific's transcontinental running across central BC from Prince Rupert on the coast to Tête Jaune Cache near the Alberta border, spawned construction of the first sawmills in the province's northern Interior. Beginning at Prince Rupert in 1908, the line was built at a far more leisurely pace than the CPR, and by 1909 builders were still uncertain precisely where it would cross the Fraser River. Nick Clark of the Northern Development Company set up a sawmill in Quesnel to cut

> The best way to build a flume was to make a pilot road. In the old days, you had to do it with a pick and shovel and a team of horses. You'd get a rough road built so that you could get one of these steam tractors up there, and they would set up a little portable mill.
>
> They would saw that stuff right there, and make the brackets and everything and start the flume building. They would build one or two hundred feet a day with about six or eight men, and they would throw the timber in and the brackets and the different dimensions for the bents and all that, and it would float down. It had a gate in there and it would hold them. That was the best way to do it. When they first made flumes in this part of the country they used to build them for a dollar a foot.
>
> Most of it was contract. Now the one out at Boulder Creek, I was there when Jacobs was building the flume there. A.H. Green was an engineer and surveyor here [Nelson] and they built the flume coming out of Archibald Creek out here by Meadows. There was about four or five miles of flume there.
>
> Most logs at that time were all standard lengths like twelve, fourteen, sixteen, eighteen. Well, they would take quite a curve. It was the cedar poles that were the hard ones to flume. They put them in with the top down so the water was working on the butt. They would ride up on the side of the flume and sometimes they didn't fall back into the flume, they fell on the outside. But most of those kind of curves, they rectified them by making them more of a compound curve.
>
> The flume at Clearwater was for lumber. They used to make a flume with a two-by-twelve at the bottom and two-by-twelves on the sides. The lumber would go floating down there pretty good. Didn't take so much water neither.
>
> There was an old-time mill in the early days at Crescent Valley. That was the Patricks, the ones that made their name in hockey. They had a big mill there and logged up the Little Slocan, and they used to drive the river down to Crescent Valley. Well, down at the lower end, a row of piling ran down the middle of the river. Down one side all the brush and stuff that was no good went, and the logs went into the other channel.
>
> They didn't have it boomed up strong enough to hold them there. When you've got, say, fifteen million board feet of logs pressing against a boom without a safety boom, if the water comes up and it gives her a flush, she breaks. You lose them all. That's what happened. They all went down into the States, the whole works, and that shut her down. Right into the Kootenay and down the Columbia. Fifteen million in one bang.
> —Russell Fletcher

lumber for a sternwheeler that would carry freight along the Fraser, connecting with the railway at Fort George. After building the 70-foot *Nechacco*, Clark loaded his sawmill aboard, steamed up the river and set up Prince George's first lumber operation. The following year he built Northern Development's second mill in the new town. He obtained logs for both mills from land clearing operations that were underway as settlers feverishly developed their townsite. A third mill, Bogue and Brown's, appeared on the banks of the Nechako River a few weeks later. All of these mills cut rough lumber for customers who stood waiting for the boards to come off the saws. It was an unspectacular beginning for what would become the Interior's leading centre of forest industry activity.

Perhaps the most elaborate combined sawmill and woods operation built at this time was at Enderby. The S.C. Smith mill, moved there in 1894, had continued to operate past the turn of the century, undergoing several expansions and changes of ownership. In 1905 it was sold to the A.R. Rogers Lumber Company of Minneapolis for $350,000. Rogers remodelled the mill, adding new planers and a lath mill, and boosting production to 150,000 feet a day. The mill also had a box factory, one of several that appeared in response to the growth of the Okanagan orchard industry. Each fruit required a certain size and shape of box, and Rogers' workers resawed lumber into various dimensions for the different parts of the boxes. Some

Nick Clark's sawmill, the first mill in Prince George, 1910.
BCARS 27519

factories assembled complete boxes, all by hand, while others sold components to orchardists.

The Rogers mill was a much-admired facility, with its neatly tended grounds and its new buildings painted bright red with white trim. Annual production eventually reached 18 million feet. The mill's timber supply lay up the Shuswap River in extensive Timber Berths. Logs were also purchased from settlers in the valley. Most of the logging in this period was done with horses in the winter, with logs, ties and poles skidded to the Shuswap and piled on the ice or decked along the riverbanks. The mill, and the thriving economy of Enderby, were entirely dependent on the annual log drive.

> There was river rights granted to various mills. The major mill, as far as I can remember, was Armstrong Sawmill. He always had the first rights on the river, which was very important for economy. The best and cheapest times to drive logs was when the water was barely up. It had risen some, and it was rising an inch or two every day. Every jam you come to, everything was loose, and it went along much easier. The second or third guy along, and the river dropping two or three inches a day, by the time he gets to his logs they are sitting up high and dry. So that was really important, to have first rights. My family was involved in river driving for at least sixty years, and I did hear a lot about it.
> —*Robert Dale, who was a logger and river driver. His family settled at Mara Lake in 1897 and has been active in the forest industry around Enderby for a century.*

The forest economy in the Enderby–Shuswap region at this time owed its vitality to a unique set of circumstances. First, there was a good supply of high-grade timber—fir, pine, cedar, larch—growing right up to the headwaters of the Shuswap, deep in the Monashees. The river itself was well suited to driving logs, beginning at Sugar Lake, then along the upper Shuswap for 25 miles (40 km) or more to Mabel Lake. Eleven miles (17.5 km) down the lake, the lower Shuswap ran another 25 miles (40 km) to the mill at Enderby.

Good farmland lay in the broad river valley between Mabel Lake and Enderby, and with the construction of the railway many settlers were drawn to the area from eastern Canada.

The Rogers Lumber office at its mill in Enderby. EDMS 2259

Several of these were experienced loggers and river drivers from Ontario, who were accustomed to farming in the summer, logging in the winter and driving logs down rivers at breakup in the spring. Perhaps more than anyplace else in BC, the Shuswap valley resembled a classic eastern logging chance.

Jobs on the river drive were much sought after. The work paid $3 a day and free board, seven days a week for the six to eight weeks it took to move all the logs. The Rogers company hired a thirty-five-man crew, who worked under experienced foremen such as Angus Woods. Some of the crew were old, experienced river hogs such as the Dale family. A few teenagers anxious to prove themselves were also hired each year.

> In the older drives, including when I was a boy with my dad, there was no such thing as a life jacket. I mean, a driver didn't fall, and if he did he had a pike pole and he could grab something or do something. Probably one of the most famous there was Henry Simard, who did a lot of river driving. Everybody swore he couldn't swim. They said he could walk on the bottom fast. My uncles, I never saw them swim. That would be an insult. You worked above the water, you didn't want to get in it, I guess.
> —Robert Dale

The first task was to pack peaveys, pike poles, batteaus, tents, food, a mobile camp kitchen and other equipment into wagons, and to make the three-day trip through Vernon and Lumby to Cherryville, where most drives began. Then the drive started. The workers rolled the decked logs into the river, which was swollen by the spring thaw. As the logs were swept downstream, the drivers followed on foot along the banks and in their batteaus—22-foot boats pointed on both ends, each with a crew of six to eight men. For ten to twelve hours a day they laboured, rolling logs off sandbars, prying them off rocks and breaking up jams. The latter was a risky task. The workers found the key log at the front end of the jam and pried it loose; as the entire jam began to move, the drivers had to pull out fast, running across the churning logs to avoid being rolled beneath them and crushed or drowned.

River driving was wet, cold work. Just getting from the river to the cook wagon and that night's camping spot sometimes required wading through back channels and swamps—as did the return to work the next morning.

After two or three weeks the upper Shuswap would be cleared, with the logs stowed in a

River driving was wet and often dangerous work.
Top: Guiding logs downstream on the Shuswap River.
EDMS 1417
Centre: Driving logs down the Slocan River to the Patrick Lumber mill at Crescent Valley.
BCARS C-09790
Bottom: Clearing a jam at the Skookumchuck on the Shuswap River. EDMS 541

bag boom on Mabel Lake. The crew then divided, most of them following a trail around the lake to the lower Shuswap, where logs skidded by settlers were put in the water. The rest of the crew moved the boom across the lake. Until a steam tug was hauled into the lake with horses and a sleigh in 1910, the workers moved the boom with an ingenious device called a horsejack. It consisted of a float about 40 feet (12 m) square, on which was mounted an iron capstan. The float was tied to the front of the log boom. A large anchor on the end of a mile-long, 1½-inch (4 cm) rope was carried down the lake in a batteau and dropped. Two horses were kept in a shed on the float and worked in shifts, one horse at a time walking in a circle, turning the capstan and pulling the float and boom down the lake. Although the horsejack operated around the clock, it was a long, slow process, taking several days.

At the outlet into the lower Shuswap, the logs were fed into the stream and down through the treacherous Skookumchuck rapids, where men were positioned with pike poles and peaveys to try to prevent jams. They were not always successful, and some of the most difficult jams occurred in this short stretch of river. One year when a large jam formed and the drive boss was unable to find the key log, he sent for some blasting powder. Most of the logs in the jam were larch, covered with pitch. When the old-style black powder was set off, the usual cloud of black smoke obscured the jam; then suddenly it was swept away in the updraft of an enormous fire. The powder had ignited the pitch and to the drive boss's horror a huge mass of logs was going up in flames. Then the jam let go, the logs slid into the current and the flames were snuffed out in an instant.

From the rapids on, jams occurred frequently in the slow-flowing river along the valley bottom. The crew spent long days waist-deep in the cold water, and became exhausted after weeks of unrelieved labour. Finally the logs were all stowed in the millpond on a back channel, and the drive itself was over. But this was not the end of the annual ritual.

> These log drives for Armstrong Sawmill, even when I was a kid, the farmers up there, everybody had a bit of wood. During the winters everybody would take some logs out, make a deck. And these decks were all along the riverbank. One of the scalers would go out and you were paid for them on the riverbank. The rule was that logs had to be scaled before they were dumped in the water. This was a good thing for the guys that were selling them, and it was also to pay the stumpage and royalties. They had to be paid before they went into the river. Then, if it was lost, it was your loss. The log drivers had these huge decks of logs and as they came down the river they had to throw these decks in. It was pretty hard and dangerous work.
>
> When the mill hired my uncles to keep the islands open, to get there they would just hop on a log and go down to the islands. They had to walk home, but they never had to walk there. They were pretty nifty on their feet. Naturally, they got a log that had a little size to it and a little peak, twist to the butt, so it didn't want to roll.
> —Robert Dale

A log rolling competition, one of the events at the celebration in Enderby following the annual log drive down the Shuswap River. EDMS 286

A successful river drive of 16 or 18 million feet of logs was crucial to the economic life of the valley and the town of Enderby. It meant the farmers and independent loggers would be paid for their timber; it meant the millworkers could expect a full year's employment; it meant local merchants would enjoy brisk sales and the entire region would prosper. This was reason for celebration, and for two or three days after a drive, the town of Enderby erupted in a festival of dances, parties, sporting events and general merrymaking. Among the highlights of this annual event were loggers' competitions—log birling, greased pole climbing, axe throwing and so on. These were the first organized loggers' games in BC.

This closely integrated pattern of logging, river driving and sawmilling on the Shuswap lasted for fifteen years, when Rogers suddenly ceased operations and closed the mill in 1922. Ever since, the reason for his apparently precipitate decision has been somewhat of a mystery in the area. Perhaps he felt the timber supply was running low, or future markets looked uncertain, or the anti-American opinions of some residents and workers offended him. Whatever the reason, he left abruptly. The mill was eventually reopened and operated on a reduced scale, and the pole business picked up some of the slack for loggers. Driving the Shuswap River continued on a much smaller scale into the 1950s. But the heady days of the big river drives and the full-tilt operation of the Rogers mill were over—well before the forest industry in most of the Interior was underway.

By 1909, with both the Coast and Interior forest industries growing by leaps and bounds, the public grew increasingly alarmed about the state of the province's forests. Opposition politicians claimed they were being overcut and predicted that the province would soon run out of timber. Because of the timber staking frenzy precipitated by the McBride government's Special Timber Licences, 9 million acres (3.6 million ha)—comprising the best and most accessible timber land in the province—had come under the control of speculators and lumber companies. Railway construction had opened the Interior to development, and the subsequent construction of a few large sawmills had established an industrial foundation for an Interior economy. But now its future began to look uncertain.

Railway building activity had started an untold number of forest fires which burned hundreds of thousands of acres of the most valuable timber in the province. Initial surveys along the upper Fraser River, through which the Grand Trunk Railway would run, raised fears about the potential destruction of a huge, valuable and, at times, tinder-dry spruce forest. Whether there would be a forest upon which to base an industry was a much-debated question.

As well, the provincial government had become dependent on forest revenues in the form of rentals and royalties, which grew from $142,000 in 1900 to $2.5 million in 1910. Significantly, in 1910 the government spent only $91,000 on forests, half of it fighting fires.

In response to this uncertainty, the government appointed a Royal Commission of Inquiry on Timber and Forestry in 1909, with former Minister of Lands Fred Fulton in charge. After a year of deliberations and public hearings, including sessions in Kamloops, Vernon, Revelstoke, Nelson, Cranbrook, Fernie and Grand Forks, the commission delivered its report.

The report was clearly inspired by conservationist principles, coming down firmly on the side of a regulatory system to protect the forests and provide a perpetual timber supply for the forest industry. It described a widespread opinion:

At the outset your commissioners were met by the argument that logging operations in this Province were in the hands of practical business men engaged in supplying a certain market demand for lumber; that the inference was that the forest was receiving the best treatment that was commercially possible; that it was absurd to suppose that any timber that could be profitably removed would be left to burn or rot upon cut-over lands; and that any arbitrary regulation of cutting by the government would be improper interference with the natural trade conditions of the lumbering industry.

The commission rejected this laissez-faire attitude and recommended a combination of higher stumpage rates and utilization regulations to reduce waste. All trees over 14 inches (35 cm) in diameter should be cut, and every tree should be utilized to a 10-inch (25 cm) top. Stump heights should be kept low, and sound dead timber used. As well, forest protection should be stepped up and a fire prevention system put in place. In short, the forests should be treated as a valuable resource and their productivity maintained.

In 1912 the government responded with a new Forest Act, establishing a Forest Service and incorporating many of the commission's regulatory recommendations. A young Ontario forester, H.R. MacMillan, was appointed the province's first chief forester. One of his first moves was to set a policy of developing an economically sound forest industry. A conservationist dedicated to the idea of a permanent forest, MacMillan understood that the future welfare of the province's forests depended upon the existence of an industry which needed and valued them. MacMillan later played a major industrial role on the Coast, founding a company which evolved into MacMillan Bloedel, until recently the largest forest company in Canada. The policies and practices he put in place beginning in 1912 insured the continued development of the industry for the next twenty years.

CHAPTER 4

Boom Years

The 1910s

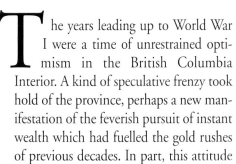

The years leading up to World War I were a time of unrestrained optimism in the British Columbia Interior. A kind of speculative frenzy took hold of the province, perhaps a new manifestation of the feverish pursuit of instant wealth which had fuelled the gold rushes of previous decades. In part, this attitude was spawned by the second wave of railway-building projects, but to some extent these projects themselves were products of the prevailing spirit of optimism. The same can be said of the provincial government of the day, Richard McBride's Conservative administration.

McBride was, to put it mildly, enthusiastic about economic development in BC. The Special Timber Licence he introduced in 1905 is one example of his style of government: it fed speculators, swelled provincial coffers and touched off a boom in the forest industry. This encouraged McBride and his colleagues even further. One result was the government's enthusiastic support of a third transcontinental line, the Canadian Northern Railway—including generous financial backing for the BC portion. This line, like the Grand Trunk Pacific, would run through the Yellowhead Pass, swing south down the North Thompson River to Kamloops, then follow essentially the same route as the CPR to Vancouver.

When construction began on these lines in 1906, astonishing speculative booms occurred along their routes. Real estate developers obtained land from the government for "a dollar and a drink," as it was described at the time, and along with the railway companies which had received land grants, sold residential and industrial lots in lavishly described towns and cities, most of which were never built. Perhaps the greatest concentration of these promotions took place around the junction of the Nechako and Fraser rivers, where the Grand Trunk Pacific was expected to cross the Fraser. In 1910, land at this location, the future city of Prince George, was selling for $10,000 an acre. McBride and his government got so caught up in their own propaganda they announced plans for a third new railway that would connect Vancouver with the Grand Trunk Pacific at Fort George, then head off to vaguely described destinations in the north. It was called the Pacific Great Eastern Railway, and after its promoters ran out of money building a line from Squamish into the Cariboo—"from nowhere to nowhere," in the language of its critics—the government took it over. In the 1910 general election, McBride's Conservatives took every seat but two. Clearly McBride's outlook reflected the spirit of the day.

Beneath the surface, however, all was not well. The mining industry was in trouble, as was agriculture. The financial sector was about to be traumatized by the collapse of Dominion Trust Company in Vancouver. Labour unrest was mounting, spurred on by intolerable working conditions in most of the province's mines and railway construction camps, and by political groups such as the Industrial Workers of the World (the Wobblies). In 1910, the year the BC Federation of Labour was formed, twice as many days were lost to strikes as had been lost in the previous five years combined. The province was polarized into hostile camps of capital and labour.

The one economic bright spot was the forest industry. As Special Timber Licences moved from the hands of speculators to investors, US and eastern Canadian capital began to flow into the province, primarily the Coast and the southeastern Interior. The prairie market was booming to the extent that BC Interior producers could not meet the demand, and coastal mills began shipping lumber east on the CPR. The promise that the Panama Canal would open in 1914 had coastal lumbermen dreaming of European markets, and their Interior colleagues looking forward to the day when they would have the prairie market to themselves.

Page 60: A huge sleighload of logs on their way to the Adolph Lumber mill near Elko, 1908. BCARS 52575

Page 61: A Staples mill logging crew with one of the company's Shay locomotives in the background. FSM F.S.5-155

> In the fall of 1910 I went back to the woods and worked for Bill Hunter—he was the MP in the McBride government and was later defeated by Sid Leary who ran against him. I was logging opposite Silverton on the mountain, where Hunter owned three or four timber licences. I used to work for Hunter in winter as a foreman, and in the summer, when his camps closed down, I got a job as foreman on the roads with the Public Works. At the Summit Lake camp there were wonderful cedar poles, some hemlock and spruce, and balsam and tamarack. It was a timber licence, but we never kept track of returns to the government then. There was an old fellow in Kaslo, Carney. He had long grey whiskers when I knew him. He used to come around every six months and ask us how much we cut, and collect the dues; and I suppose turned them in.
> —R.E. Allen, a logger, mill operator, forest officer and district forester throughout the Interior, beginning in the early 1890s.

The combination of brisk lumber markets and the BC government's promotional hype generated a dramatic expansion of the forest industry in the southeastern Interior. Most of this development took place in the east Kootenay region, centred around Cranbrook and Fernie. Several dozen new sawmills were built in the eight or ten years before the war. Many of them were large for their time, and many had extensive logging operations that employed steam-powered equipment.

Until about 1910, most of the mills in this region had been built in and around Cranbrook and Fernie, and some had been operating since before 1900. Now, as business picked up, the immediate timber supply was soon exhausted and new mills were built farther from these centres. Towns and villages grew up around the mills, spreading the industry throughout the region.

Fernie, which was closest to the prairie market, went through the expansionary phase first. The process of dispersal was hastened by a massive fire in 1908, which burned the entire city of Fernie and a vast area of surrounding timber. Sixteen miles (26 km) to the south, at Elko, Elk River Lumber built a 50,000-foot-a-day sawmill. It included a planer, and the mill finished lumber for several smaller sawmills in the area. A mill of similar size was built by Crows Nest Pass Lumber at Manistee, 12 miles (19 km) south of Elko. A couple of miles away, at Waldo, two large mills belonging to Baker Lumber (which started life in 1909 as Wood &

Top: Baker Lumber's double-cut band saw at Waldo was the only one of its kind in the southern Interior in 1910. FSM 8.559

Bottom: The Ross-Saskatoon Lumber Company's Shay locomotive hauling logs through a cut-over area near Waldo, 1909. FSM F.S.9-4

McNab Lumber) and the Ross-Saskatoon Lumber Company each cut 75,000 feet a day. Nearby, at Baynes Lake, Adolph Lumber had a mill of equal capacity, around which a town of 250 people grew. Both used steam locomotives running on narrow gauge rail lines. Beyond that, at Flagstone, a 25,000-foot-a-day mill operated for a few years, supporting a small community with a hotel, school, store and post office. At Hanbury, 30 miles (48 km) east of Fernie, another 50,000-foot mill supported another community of 100 residents.

As forests in the immediate vicinity of Cranbrook were cut, other communities grew up around mills in outlying areas. To the southeast, at Jaffray, two mills—the East Kootenay Lumber Company, built in 1907, and Jewel Lumber, started in 1909—cut a combined volume of 120,000 feet a day. East Kootenay was a railway operation with 10 miles (16 km) of track and a single geared locomotive. A few miles closer to Cranbrook, at Wardner, Crows Nest Pass Lumber had a sawmill which could cut 100,000 feet a day. This company ran mills at several locations and shipped logs where they were needed, through its own extensive railway logging operations, which tied in to CPR lines.

*The Staples mill near Wycliffe, shown here in **1903**, transported its logs on the longest logging railway in the Interior.* FSM F.S.5-153

Southwest of Cranbrook, along the Moyie River, several other lumber communities flourished at this time. Two mills at Kitchener together cut 180,000 feet a day. To the west, at Wynndel, Monrad Wigen and his father started a steam-powered box factory in 1913 to supply local fruit growers. Wigen cut the lumber on a portable mill with which he cut railway ties in the winter. This company evolved into Wynndel Box & Lumber, which is still operated by its founder's grandsons.

In 1910, A.E. Watts, an American lumberman, built a mill about 10 miles (16 km) southwest of Cranbrook. A somewhat contentious individual, Watts insisted that the community which grew up at his mill on the CPR line be called Wattsburg. The CPR disagreed, calling their station Lumberton. This was the name adopted by the Wisconsin-based BC Spruce Mills Ltd. when it bought the mill a few years later and turned it into the largest sawmill in the Interior.

North of Cranbrook, near Fort Steele, an eastern investor named Jonathan Bridges bought 30 million feet of timber on CPR railway-grant land for $30,000 to launch Bridges Lumber. He set up a sawmill and logging operations, and spent the next ten to twelve years enduring a long series of misadventures and misfortunes which involved most other mill owners in the area. Finally he went bankrupt.

Nearby, at Wycliffe, Otis Staples built a 100,000-foot-a-day mill whose timber was supplied by the most extensive railway logging show in the Interior, an operation chartered in 1906 as the St. Mary's and Cherry Creek Railway, but which never functioned as a public carrier. It had 14 miles (22 km) of track and three Shay locomotives. Staples also used three steam-powered donkey engines to skid logs along the ground. He may have been the only logger to do so in the Interior at this time, although the method was commonly used on the Coast and in parts of the USA.

In 1913 R.E. Benedict and Lafon were borrowed from the US Forest Service to organize the BC Forest Service. Benedict came along and asked me how I would like to go to Hazelton as district forester in the newly organized service. I wasn't too keen on it, but as it was a promotion, I agreed. I went down to Victoria first and got my instructions from MacMillan, Lafon and Benedict, and then moved to Hazelton.

I took the job as district forester at $114 a month—that's less than I got as fire warden. I got $1 a day for an old cayuse that I bought for $5, and I got $2.50 a month for shoeing him; and I got $20 a month living allowance for being out in the wilderness. So, all told, it amounted to about the same pay.

The first year that I was there they called us down to Victoria to the first district foresters' meeting. I told MacMillan that I couldn't support my family on the salary I was paid. MacMillan said that if I did as well the following year as I had done, he would pay me $1,800 a year, so in 1915 I got my $1,800.
—R.E. Allen

Top: Loading with a steam jammer mounted on a rail car at Staples' Wycliffe logging operation. This machine was probably built in the company's machine shop. FSM F.S.5-156

Bottom: Workers crossing the Bull River in an aerial sling, 1920. FSM 18.5

Another extensive logging and sawmill complex, this one without a railway, was built by the CPR on the Bull River, southeast of Cranbrook. After completion of its main and branch lines, the railway had become a voracious consumer of timber for use in stations, warehouses, boxcars, telegraph cross arms and a huge volume of ties, which had to be replaced regularly. To meet these needs the CPR's Tie & Timber Branch had set up its own logging and sawmill operations, beginning with a mill at Fernie in 1899. This mill was destroyed in the big fire of 1908 and was relocated to a site nearer a good timber supply, on the Bull River where it joined the Kootenay. Construction began in the summer of 1911, and that winter full-scale logging, employing between 500 and 600 men, began at a series of camps established up the Bull. It was the most extensive logging operation in the Interior, running almost fifty camps altogether. The annual harvest during the early years was as high as 25 million feet of sawlogs and a million ties. All of this was accomplished by the labour of men and horses, as building a railway into the rugged Bull valley was impractical, and no other mechanized equipment was available.

The CPR employed its own workforce to build the main camps and roads, and the chutes and flumes. Logging was contracted out, some to full-phase contractors who delivered their logs to the river, and some to smaller companies which worked under CPR supervision to carry out contracts for falling and bucking or skidding and decking. The company maintained a large supply camp at the mouth of Galbraith Creek where contractors could rent all forms of logging equipment, from axes and saws to horses, harness and sleighs. The supply camp alone had a barn for eighty horses, and there were smaller barns at the other camps. A local veterinarian regularly examined and cared for the horses.

The Bull River logging system was a classic horse logging operation of the late nineteenth and early twentieth centuries. The methods employed here had been developed in Nova

Scotia and Maine, then perfected in the woods along the Ottawa River and in Wisconsin and Minnesota. The loggers too had migrated across the country, working their way west from camp to camp. Their ranks were augmented by prairie farmers—many with experience logging in the east or in Europe—who worked in logging camps in the winter to finance their farms. Backed by the financial resources and organizational expertise of the CPR, the Bull River logging complex was the epitome of premechanized timber harvesting for almost twenty years.

There was big timber on the Bull, mostly ponderosa pine and Douglas fir, up to 7 feet (2 m) in diameter. Fallers used a springboard—a 5-or 6-foot (1.5–2 m) two-by-six with a horseshoe nailed on one end—which they inserted into a notch cut in the tree to get above the swelled butt. Crews of two or occasionally three men felled, limbed and bucked the trees into about 10,000 feet of logs a day, for the then-princely sum of $1 to $1.50 per thousand. The actual job of felling a tree began with a horizontal saw cut facing the direction of fall, made with a two-man saw that had a handle on each end (for a long time the favourite saw in the Interior was a 5½-foot-long Disston Royal Chinook). A wedge of wood was removed above this cut with axes. Fallers favoured double-bitted axes, with one blade honed to a razor edge for falling, and the other to a blunter finish for limbing. The shaping of this undercut, as it was called, determined precisely the direction in which the tree would fall. Then the saw was used to make a falling cut in the back of the tree, 2 or 3 inches (5–7 cm) above the undercut, to prevent the tree from jumping back at the sawyers when it toppled.

Once the top was removed and the limbs were cut off flush with the trunk, the tree was bucked into lengths of 16 to 24 feet (5–7 m), depending on its diameter and the difficulty of the ground over which it had to be skidded. In deep, soft snow trees would have to be dug out so they could be bucked and skidded. After they were bucked, the logs were rolled with a peavey and the limbs on the bottom side were cut off. Big logs might be "sniped," or bevelled on the leading end to prevent hangups when skidding.

At a well-co-ordinated logging show like the one at Bull River, skidding was usually done in two phases. In the first, a swamper cut a rough trail from the bucked logs to the nearest skid road. Then the teamster, with a team of compact, nimble horses, dragged the log out through the stumps and brush. At the skid road, another teamster with a larger, heavier team of horses added the log to a string of perhaps a dozen similar logs attached with a short chain and a dog on each end which was driven into the logs. On longer hauls, and over suitable terrain, the logs might be loaded on a sleigh, or a sloop if the route went down a steep slope. Logs were loaded on sleighs and sloops with peaveys from a deck above the sleigh road,

The CPR's main logging camp on the Bull River at Galbraith Creek. BCARS 73470

or by cross-hauling them using a block in a nearby tree, and a line with a set of hooks on one end and a horse hauling on the other.

There was a special method of securing logs to a sleigh or sloop. The two bottom, outside logs were tied to the sleigh bunk with special chains called corner binds. Then the space between them was filled with a layer of logs. A chain was wrapped around this layer and another layer was loaded on top, resting on the wrapper chain and pulling the load tight. Five or six layers were stacked up in this manner, making huge loads.

The sleigh or sloop was pulled along a specially prepared road, the snow ploughed regularly and the surface packed and occasionally sprayed with water to maintain an icy surface. To slow down the device on a steep slope, various means were employed. "Rough locks" were chains tied around the rear runners at the top of a hill; hay or sand was spread on the downhill portion of roads; in some cases, a device called a hill brake or crazy wheel was used. It consisted of a set of four to six wheels, each with a brake pad which could be applied with a long lever. A cable or rope was threaded around the wheels in an intricate pattern and attached to the rear of the sleigh. The brake was chained to a tree or stump at the top of the hill. By gradually easing off on his lever, the hill-brake operator could control the speed of the sleigh as it ran down the slope.

> In the early days we had no motive power other than horses, and when they came to an adverse grade they just stopped. On a good road and for a short haul, a four-horse team with sleighs on snow or ice might move two thousand feet up a 4-percent grade. After going about two hundred feet you would have to stop and "wind" the horses and it would be very likely that they would not or could not start again, so you would have to either lighten the load or get more horses.
>
> When Dad and I were logging at Jura, the timber was back a few miles from the railway track, and mostly all hauled out on snow or ice roads in winter by horses with sleighs. Here we could haul up to three thousand feet per load. Some of the hills were quite steep, and the downgrade had to be sanded. We kept one or two sandhill men on the roads at all times. An old southerner by the name of Black Sam worked for us and drove his own outfit. One snowy morning I could see that the sanders had missed some spots and I asked Sam how he came down. "Right smart!" was his answer. This was in 1926 and just about the end of hauling with horses since trucks and Cats came into use then. We used a light Cat for the next two years. In the winter we used some of the old horse sleighs and in summer we changed to steel trucks with twelve-inch tires.
>
> We had used these steel trucks when logging for the old Canyon City Lumber Company of Creston, about ten years before coming to the Okanagan from the Kootenays where my father logged the townsite that is now Creston. Canyon City Lumber also owned some of the big Michigan wheels, and when Hugh Leir heard of these, he sent my father over to Creston to get the steel trucks and two sets of the Michigan wheels, one for the company and one for us.
>
> These big wheels were drawn by horses and had a slip tongue arrangement working on a half moon on the axle. In the woods you drove over and straddled the load, then put the chain around the logs ahead of centre. This chain had to be placed just right from centre depending on how steep the grade of the road was. In fact, the advantage of these wheels was that not much of a road was required since they would roll over just about anything. However, the pole would whip back and forth and could kill you or the horse if you didn't stay clear; after a few hard knocks you did. The system was that you rode one of the horses and as the team and pole moved ahead, a chain on a short lever rolled the half moons. This either lifted the logs clear of the ground or let the back ends drag and act as a brake. Then on a steep pitch with the horses not pulling, the pole would slip back and the whole load would drag. One of these old sets of wheels is in Princeton, and the other is in Penticton.
>
> —*Jack Broderick, who began logging about 1920 near Penticton and spent the rest of his life logging in southern BC.*

A big wheel with the logs raised under the axle at Canyon, 1910. BCARS 56351

When braking devices were not used, or when they failed, the teamster had to drive his horses at full gallop to keep ahead of the out-of-control load. It was not a particularly dangerous task for him: he sat on top of the log load and he could always jump off if disaster

Different conditions of climate and terrain called for different horse logging systems.

Clockwise from top: Loading a sleigh at a Columbia River Lumber show on Horse Thief Creek, 1910 (BCARS 60229); **skidding logs from the stump to a skid road, using tongs to hold the log, near Cranbrook, 1910** (CFI); **a loaded sleigh being eased down a hill with the help of a hill brake, BC Spruce Mills' Lumberton operation, 1922** (BCARS 73599); **a sleighload of logs being brought into a BC Spruce Mills camp near Lumberton, with hay strewn on the road to help with braking, 1922** (BCARS 73571).

was imminent. For the horses it was different: when they were overtaken by their load, they were usually killed. In a logging operation of this sort, teamsters were the most skilled and, along with fallers, the highest-paid workers. For the most part their job was a cold one, riding through the frozen forest atop a load of logs and unable to move around much to keep themselves warm. The teamster equipped himself with a "dry ass," a gunny sack filled with hay, to sit on.

The logs were hauled to the Bull River, or sometimes to loading ramps at chutes and flumes, in which logs were moved down particularly steep grades to landings or to the river. Along some of the creeks, flumes were built and logs decked at their heads until spring. The flume down Tanglefoot Creek was more than 4 miles (6 km) long and was used to float logs and ties. Splash dams were built on some of the creeks to flush logs to the river.

At the river, the logs were rolled off the sleighs or sloops and cross-hauled or rolled with peaveys into decks to await the thaw. Ties were loaded by hand, laid crosswise on the sleighs, hauled to the river and decked in huge piles of as many as 100,000.

In early spring, with the ice gone out of the Bull, the log drive down the river commenced. The hundred or so men employed to run it used traditional river driving methods, but they had to deal with one of the wildest, roughest log rivers on the continent. One experienced French Canadian crew brought in for a spring drive were stymied by the river. Most of the work was done by loggers who lived in the area, and a few professional drivers from elsewhere. The Bull is fast and turbulent along much of its length, and in two canyons the stream narrows to less than 20 feet (6 m). Logjams were common in these canyons, and they were difficult and dangerous to break up. When log traffic in the 40 miles (64 km) of river above the canyons was heavy, huge jams formed quickly, filling the 100-foot-deep (30 m) canyons to the top. Then the only recourse was to lower men into the canyon with ropes so they could pack a charge of dynamite deep under the logs. On one such occasion, with a million and a half feet wedged in the canyons, 30 tons of dynamite was needed to clear the river, most of it exploding in one blast that scattered ties and logs a mile in every direction.

Eventually the logs and ties ended up in the millpond, behind a dam just above the sawmill. A flume carried the ties past the mill to a tie deck on a rail spur where conveyers moved them to boxcars. The sawlogs were sorted and fed up a jack ladder into the sawmill.

The mill was powered by steam boilers fired with shavings and sawdust. These boilers drove a single engine with one huge piston and a 12-foot (3.5 m) flywheel, located under the mill. A flat belt from this wheel transferred power to a shaft running the length of the mill floor, from which belts ran to the various mill machines.

The mill pond at the Bull River townsite, at the end of the river drive, 1915. BCARS 73471

One of these machines was a double-bladed head saw and a shotgun carriage, named for the sound it made when the steam valves released. Two men rode this carriage as it lurched back and forth under the control of the sawyer. The dogger secured the log to the carriage with a pair of tongs, or "dogs," while the setter, on a hand signal from the sawyer, adjusted the log for the next cut. From here the rough boards were fed through an edger, which cut them to width, and a trim saw, which cut them to length. They then emerged onto a conveyer, the green chain, where they were graded and removed for piling and drying before being planed.

At Bull River the planing room also contained a lath mill and a large plant for manufacturing the various products needed by the railway in its operations throughout southern BC and across the prairies—cattle guards, cross arms, grain doors, etc.

For much of its operation the Bull River mill ran two shifts, turning out 65,000 feet a day from the sawmill. It operated for seven or eight months a year, from spring thaw until the millpond froze over each fall and logs could no longer be hauled up the jack ladder. Production began to taper off in the 1920s as the timber supply dwindled. The last big log drive was held in 1927 and the mill closed entirely in 1928. The CPR opened a new mill at Canal Flats the following year.

While the Bull River operation was typical of turn-of-the-century premechanized logging shows, it was unusual for that period, when the lumber boom was sweeping through southeastern BC. Far more common were the various railway logging operations which began in this region before World War I. By this time the technology of railway logging was fully developed and ready to be imported to the BC interior. Motive power was available from several locomotive manufacturers, including small models like Columbia River Lumber's Countess of Dufferin, a Baldwin rod engine.

Loading logs with a Barnhard loader at Columbia River Lumber. The loader winched itself along a set of tracks on top of the rail cars.
BCARS 73494

This kind of engine was poorly suited to railway logging. Designed to operate on relatively flat main-line railroads, it lacked the pulling power required for the steep grades found in most logging sites, and was too long to manoeuvre in the tight quarters of most operations. In the late 1800s, three US companies began producing gear-driven locomotives designed for logging. A Michigan logger named Ephraim Shay developed a compact, powerful locomotive with vertical steam cylinders on one side of the engine, and power transferred to the wheels through a geared drive shaft. It first appeared in 1881, and by 1900 more than 500 Shays were operating in the woods of North America.

In 1888, the Climax locomotive was first produced in Pennsylvania. Its two cylinders were mounted at an angle, one on each side of the boiler, which turned a crosswise drive shaft that transferred power to the wheels via a set of bevel gears. The third logging locomotive, the Heisler, also appeared in Pennsylvania, in 1894. Its cylinders were angled down beneath the boiler and drove a central drive shaft which powered the wheels through sets of gears. Although Heislers were the least popular of the gear engines, Columbia River Lumber acquired three of them for use at Golden. Shays were the most popular, and were used by a majority of early Interior railway loggers.

Following pages:
One of Columbia River Lumber's three Heisler locomotives with a full load of logs backing down the line near Golden.
BCARS 73491

Most of these railway shows were located in the east Kootenays. There were good markets for the types of timber growing there and the relatively flat terrain in this region was conducive to the use of railways. East Kootenay Lumber at Jaffray, one of the smaller companies with a 50,000-foot-a-day mill, hauled logs over 4 miles (6 km) of track in 1908. Eastern BC Lumber ran an even shorter line at Cedar Valley, as did Fernie Lumber and North Star Lumber in Cranbrook. The Ross-Saskatoon Lumber Company, started by Hales Ross at Waldo in 1912, operated a narrow gauge railway show until it closed in 1923. This company had two Shays running on about 10 miles (16 km) of track. The biggest operation by far was the Otis Staples Lumber Company at Wycliffe.

Outside the Kootenays, at several locations connected with the CPR main line, several more logging railways operated around 1910. Columbia River Lumber at Golden was the oldest in the Interior. Arrow Lakes Lumber, at Galena Bay on Upper Arrow Lake (not the Lamb-Watson operation, which was later renamed Arrow Lakes Lumber), was a short-lived venture which began in 1907 and was shut down by 1912. A couple of others, Gibbons Lumber at Revelstoke and Hood Lumber at Taft, west of Revelstoke, lasted only briefly. Mundy Lumber at Three Valley Gap met the same fate. It pushed 6 miles (10 km) of track into the northern end of the Monashee Mountains in 1916 and used two Climax locomotives to haul out logs before going broke in 1918. The extensive timber stands on its timber berths turned out to consist mostly of hemlock riddled with rot.

Most of these operations used wood for fuel, as is evident in photographs from that period, which show large spark catchers on the exhaust stacks. But Nicola Valley Pine Mills, at Merritt, burned coal from nearby deposits in the three Climax locomotives it used until 1926.

> In 1917 they amalgamated Prince Rupert and Hazelton Forest Districts and moved me down to Prince Rupert as the new district forester. I was in Prince Rupert for three years until 1919, when I resigned and went into the lumber business with Olof Hanson at Hansall—"Han" for Hanson and "all" for Allen. I was postmaster, justice of the peace and anything else that was needed. I started from scratch with Hanson, and we worked the area around Hansall for six years until the timber ran out. Hanson owned a licence there, and we bought a lot of timber from Indian reserves and timber sales.
>
> We had a circular head rig, edger, trimmer and planer, but we had no kilns—all our stuff was air-dried. We cut about 20,000 a day and had about ten men in the mill. We paid about 60 cents an hour and had a ten-hour day, six days a week. We ran a bunkhouse and cookhouse and charged $1 a day for board, and broke even. We had some good cooks and had to feed about thirty-five men, because we had the bush crew too. The OBU [One Big Union] pulled a strike on us once, from May 1 to July 15, 1921.
>
> We had some good contracts from the CNR. They built a big fishery dock in Prince Rupert and also Jasper Lodge, for which we supplied most of the lumber and logs, all cut to specifications. The lodge logs were spruce, which took a good coat of varnish or shellac, and didn't check as much as cedar.
>
> Once a Swedish forester came up to our mill and I pointed out some of the small logs and asked him how much they would get in Sweden for similar logs. "Well," he says, "you'd get about ten years for cutting them."
> —R.E. Allen

In most industries the beginning of mechanization creates a need for even more machinery, and logging was no exception. Initially it was not difficult to get sufficient logs to railside to keep a locomotive busy enough to justify its purchase—more horses and teamsters could always be hired. But loading the logs posed a different sort of problem. Rolling them on with peaveys from elevated decks was not always possible, and cross-hauling with horses required that the locomotive wait and spot log cars—a time-consuming and expensive process.

Above: Cross-hauling logs onto a rail car with a horse, at a Nicola Valley Pine Mills siding in the Nicola Valley. NVMA

Right: Using a jammer to load logs onto rail cars near Jaffray, c. 1910. The two-man loading crew have attached end hooks to the log and are guiding it onto the load with ropes while the chaser waits to release the hooks after the log is in place. FSM F.S. 8-392

The next step for several of the early railway operators was to buy steam-powered log loaders, often called jammers. By 1910 several models were available in North America. The typical loader consisted of a powered winch which pulled in a cable that ran through a boom to a set of tongs or hooks, which picked up logs and swung them onto the railway cars. The most common model in the Interior was the Barnhard steam loader, built in the United States and used extensively through the southern USA at this time.

In the hands of a skilled operator, aided by two hook setters and a chaser to release them, the new machine could load a rail car in a fraction of the time it took with a cross-haul system. But the new system was dangerous. The operator paid out the winch line and the hook setters set the tongs into the top of the lot, one near each end, or attached end hooks on the ends. Then they ran for their lives while the operator reeled in the slack and hoisted the log in the air over the car. When jammers equipped with turntables were developed, logs could more easily be swung from decks beside the tracks over the rail cars and stacked on the load. If the tongs or hooks failed to release when the line was slackened, the chaser knocked them free, and the process was repeated until the car was loaded. Some of these machines travelled

A rare sight in the Interior—an early steam donkey used to ground skid logs. This one was possibly in use at a Staples Lumber logging show. The upper cable is a haulback line used to pull the mainline out to the log. The right-hand cable is a strawline used to pull the heavier mainline and haulback lines out into the setting.
FSM F.S. 124.09

along the empty cars on tracks or decks; others, such as the McGiffert loader, were mounted on their own cars, which could be raised on hydraulic jacks and empty log cars pulled beneath them. A further advantage of the jammers over previous loading systems was that the cars could be equipped with stakes over which the loader could lift the logs, making much larger payloads possible.

At this stage, little else in the way of mechanization was occurring. Otis Staples made some use of steam-powered cable yarding machinery of the type used on the Coast and in southern US swamps, but no other Interior companies appear to have mechanized this phase of logging. The die was cast, however, and although horses would still be used for another forty to fifty years in the Interior, mechanization of logging was underway.

By expanding rapidly during the prewar period, the sawmill industry in southeastern BC sowed the seeds of its own demise. The number of large mills built was out of proportion to real market conditions, and their capacity far exceeded the timber supply they were designed to process. The weaknesses of the industry began to show in the early years of World War I. In the Cranbrook area, for instance, combined production of sawmills in 1913 was 148 million feet; it fell off somewhat the following year, and in 1915 it was only 75 million feet. The response of most operators was to continue cutting and to lower prices.

Almost all the mills survived this first brush with disaster and recouped their losses when markets improved in 1916. They rejoiced through the postwar boom and into the 1920s. But by the end of the decade, almost all of them had closed. The collapse could not be blamed on a decline in sales or prices, and it preceded the Great Depression. Most mills had simply run out of timber, for fairly simple reasons.

First, there were no restrictions on the amount of timber harvested. The government did not

> Prior to 1913 there had been some years of stringency and of unsatisfactory market conditions in the lumber business, not alone in the Cranbrook district, but throughout the entire area embraced by the operations of the Mountain Lumber Manufacturers Association, at the annual meeting of which organization, held at Nelson on January 12, 1914, the president, W.A. Anstie, stated that "1913 has proven possibly the worst of a succession of disappointingly unprofitable years for the lumber industry of the mountains. Apart entirely from the fact that the world-wide financial conditions have been so stringent as to result in a tremendous check to the rapid development of the western Canadian cities and towns, it has been forced to the conviction of observant lumbermen that the industry has been over-exploited to an unfortunate extent. So much so that it can be said with truth that the overproducing capacity of our sawmills is out of all proportion to the existing market conditions."
> —from the Cranbrook Herald, January 20, 1916.

regulate annual harvests and the industry made no attempt, individually or collectively, to restrain itself. The only concern with overproduction, as expressed by the Kootenay-based Mountain Lumber Manufacturer's Association, was that it might lower prices. No one seemed worried about maintaining the timber supply. The industry, in spite of the conservationist sentiments of many of its members, was founded upon the old principle of "cut and get out." To compound the problem, the mills of this era, in this region, were designed to cut a fairly narrow range of the available timber supply—primarily large-diameter Douglas fir and ponderosa pine. But a limited amount of this timber was available and, given the utilization standards and forest practices of the day, most of the mills had consumed their supply within fifteen or twenty years.

Second, fire destroyed many acres of forest land. Although much of the southeastern portion of the province lies in the Interior wet belt, it is still prone to fire during summer months. The high natural incidence of lightning fires in this area was multiplied many times over by fires started by settlers and miners—and, ironically, by the loggers themselves. Crews who were inexperienced and sometimes careless in the use of the new steam-powered, wood-fired logging machinery, mostly locomotives, were probably responsible for most of the forest fires in this time and place.

An interior view of the Columbia River Lumber mill at Golden, showing the head saw carriage with the setters and dogger waiting for a signal from the sawyer. GDHS PO 559

In 1918, the federal Commission of Conservation issued an exhaustive report, *The Forests of British Columbia*, which documented fire losses up to that time. Describing the province as a whole, it stated:

> Of the forest-land, only about one-third now carries timber of commercial value, and on 97,333 square miles of forest-land, the merchantable timber has been cut or destroyed by fire. Previous to 1917, only about 30 billion feet had been cut in the province. Since most of this timber was cut on the coast and from the heavier stands, the area logged probably would not exceed 2,000 square miles. The forests of the remaining 95,333 square miles have been destroyed by fire. It is estimated that, in addition to the area on which the merchantable timber has been totally destroyed, about one-half of the area still carrying merchantable stands has been seriously damaged. It is estimated, from these figures and the average stands on unburned areas, that the amount of timber destroyed by fire in British Columbia is at least 650 billion feet, or nearly 22 times as much as has been cut by the lumbermen.

Throughout much of the Kootenay and Columbia regions, this conflagration was widely looked upon as a blessing. Areas burned or logged were much sought after as grazing land by cattlemen, who obtained large lots in the form of agricultural leases. As one rancher near Cranbrook told the province's chief forester at a later date, during a discussion on reforestation: "Evergreen trees take too much out of the soil. They ruin grassland, they poison the soil. Evergreen trees and foresters are lice on the face of the earth."

Nor would the situation change much over the next twenty years. A report prepared by Fred Mulholland, a forester, for the BC Forest Service in 1937 stated that the average annual timber harvest in the Nelson Forest District—which included all of southeastern BC—was 193,551,000 board feet between 1927 and 1936. In this same period, the average annual loss of timber to fire was 197,100,000 board feet.

One by one, practically all the large mills built in the southeast closed, and the communities which had grown around them were quickly abandoned. Towns such as Jaffray, Bull River, Wycliffe, Yahk, Kitchener, Wardner, Lumberton, Elko, Manistee, Flagstone, Hanbury, Waldo and Baynes Lake disappeared entirely or shrank to a fraction of their former size.

The first hard lessons of forest conservation had been learned in BC. They could have been learned from the experience of others in the woods of Michigan, Minnesota, Wisconsin and southern Ontario, but they were not. Neither, however, were these lessons lost upon British Columbians. The rise and fall of the southeastern forest industry prior to 1930 played a major role in the subsequent evolution of provincial forest policy, and in the eventual adoption of sustained yield forestry twenty years later.

The Columbia River Lumber sawmill at Golden, one of the Interior's largest mills from its opening in 1900 until it closed in 1927. Its large inventory of lumber is drying in the yard (left centre). The town of Golden can be seen just behind the mill. GDHS PO 555

CHAPTER 5

Supply, Demand & Depression

The 1920s

After World War I, the British Columbia economy improved gradually. In fact, recovery began in 1916 and gained momentum after the armistice two years later. By the early years of the 1920s the province was enjoying the biggest boom ever, which lasted until the worldwide financial collapse at the end of the decade. The Interior forest industry's fortunes during this time varied, depending on location.

Some of the established mills, such as the Leir mill in Penticton and Waldie's on Lower Arrow Lake, worked right through this period and beyond. Many others, including those around Fernie and Cranbrook, closed because of timber shortages—or, in some cases, for other reasons. Adams River Lumber at Chase, which closed permanently in 1925, still had a large timber supply but folded in the face of competition from other Interior and coastal mills. Columbia River Lumber at Golden closed in 1927 after a fire. A.R. Rogers closed his mill at Enderby in 1922, for somewhat obscure reasons. These were three of the largest mills in the Interior, with ample timber supplies, yet all of them failed during a time of good and growing markets.

Each of them was virtually the only employer in its community, so three company towns were plunged into hard times for many years.

Here and there throughout the southern half of the province, a new mill was established. Even before the war, in 1913, Stanley Simpson set up a small millworking plant in Kelowna. He expanded slowly through the war, manufacturing windows and doors. In 1924, he incorporated the S.M. Simpson Company and added a box factory to his operation. He also bought David Lloyd-Jones's Kelowna Lumber Company. Later that year he put in a veneer plant to make grape and berry baskets. A couple of years later he bought a small portable sawmill, which he operated at various locations around Kelowna to cut rough lumber that was planed at the Kelowna Lumber mill. Simpson's well-founded business sold lumber primarily to the local market and boxes to the growing orchard industry.

Apart from the Simpson operation, the only other newcomer of lasting significance to the business in the Okanagan was an energetic and resourceful individual, Henry Sigalet, who opened a pole operation at Sugar Lake.

> In January 1921, I went with Fred McVicker up to Prince George where there was a pulp cruise in the offing. McVicker was in charge of all cruising activities and I went up to be in charge of the cruising party... We started on the cruise on the fifteenth of January, right in midwinter. The winters up there weren't very reliable, that is to say one year you would have extremely cold weather and another, even the swamps wouldn't freeze up. It must have been pretty cold—anywhere from twenty to forty below zero, but it was beautiful clear weather and lots of snow.
>
> It was a tough job living in camps, all in tents. Each tent had a Queen heater—a round tin stove which gave out tremendous heat—and four of us to a tent. Once the Queen heater was going, things weren't too bad. But the capacity of those stoves to heat a tent wasn't a bit of use when you were getting out of your blankets in the morning to light the thing. We took turns at that and it was an awful cold job.
>
> On the whole, however, the cruise was fun and interesting. We saw wonderful country, lots of timber—principally spruce, and that's what we wanted. The purpose was to supply a pulp mill and in that day spruce was what they thought they had to have for pulp. The mill was going to start building at Prince George in nothing flat and we just had to find the pulp timber to feed it. I think it was supposed to be a 500-ton mill. We found the timber, but Prince George is still waiting for the pulp mill.
>
> When I was assistant district forester at Nelson, my work was chiefly confined to management—everything from timber cruising, land classification, pre-emptions, inspections. Also, George Melrose brought in the US Forest Service stumpage manual to be our guide and comforter in doping out timber sale appraisals. Up until that time, a very sketchy effort was made to arrive at stumpage. I think it was what we thought the operator could pay, because at that time the operations were all small and they definitely couldn't put up a very large deposit. About all they did have in the timber sale was buying themselves a job.
>
> —L. Sawyer Hope, who started working for the BC Forest Service in 1920.

Page 80: The Munson sawmill at Winfield in 1926, just after, S.M. Simpson purchased it and moved it to Kelowna.
KLM File 28

Page 81: A crawler tractor winching in a turn of logs at McLean Lumber's Shelley logging site, early 1940s.
FFG P996.3.74

From this base he would eventually start several sawmills at numerous Interior locations.

In Nelson, two unique plants were built during the 1920s. W.W. Powell, an American who operated a mill at Spokane that supplied match blocks to eastern US match companies, set up a mill in Nelson in 1921 to cut lumber and match blocks from the white pine stands scattered through the Kootenays. Powell acquired very little of his own timber; instead he bought logs and lumber from other mills with rail or water access to Nelson. Pine of sufficient quality to produce match blocks was cut into 2-inch boards which were dried for a year or more. They were then cut into a variety of match-length blocks. A conveyer carried the blocks to a sorting table, where a crew of about twenty women chopped out defects and packed the selected material in boxcars for shipment to match companies across Canada. The plant's sawmill also cut standard dimension lumber which was sold at a premium in eastern Canada for pattern stock. The match business, being relatively stable, kept the Powell mill operating steadily for several decades, until it closed in the early 1960s.

Women inspecting match blocks as they emerge from the gang saw at the Powell mill in Nelson, c. 1925.
KMA 92.97.169

Also at Nelson, the Interior's first plywood plant opened in 1927. BC Veneer Works produced up to 6 million square feet (540,000 m^2) a year of larch, cottonwood and birch plywood from logs cut by other sawmill companies throughout the Kootenay watershed, mostly from the Lardeau district. During World War II the plant supplied birch veneer to aircraft manufacturers. Operations were suspended in 1945 because of a shortage of cottonwood, and the machinery was sold to John Bene at Western Plywood in Vancouver.

Most of the new mills built during the postwar period were farther north, along the new Grand Trunk Pacific line east of Prince George. The railway was completed in 1914. By 1919 it was broke and had been taken over by the federal government. Later it was incorporated into Canadian National Railway. In the early years of the war, several of the small mills that had been established there attempted to stay in business at various points along the line, which ran parallel to the Fraser River east to Tête Jaune Cache. Known as the East Line, this timber-rich valley contained huge stands of spruce and was fed by several major tributaries. But during the economic downturn that occurred during the first half of the war, most of these mills disappeared, except for an 80,000-foot-a-day mill built in 1913 at Hutton by Bill Willots.

Roy Spurr had come from Nova Scotia. He had had a couple of partners in with him when he first started and he bought them out. He was one of the old-timers of the district, and had run scows during construction days on the Fraser River. He had the best natural eddy and booming ground for holding logs on the river. All the weight of the logs was held against the bank, and when a boom was put along the outside it would hold a fair supply of logs all the time and none were lost. The other mills on the river had booming problems all the time. The river shoals, you see. Roy Spurr had his mill right on the river and the spur down the track. He was a good mill man and was very particular that his stuff was well sawn and well seasoned. He was also a very careful businessman. At that time it was a bad show and I used to have many a talk with him. Once, when I was up there, he had a very tragic accident in his mill. The sawyer stepped forward to check something and forgot to kick the safety device on the control to the carriage. The carriage flew forward and put him right into the saw.
—*J. Miles Gibson, who worked for the BC Forest Service from 1920 to 1930, including a stint as district forester at Prince George. From 1930 to 1962 he was dean of forestry at the University of New Brunswick.*

There was not that many sawmills up there in those days and they were pretty much all located along the line of the CN and the Fraser Valley, between McBride and Prince George. There was one at almost every siding. They were getting their logs by skidding them into the mill. As you know, one by one they kept going broke. Almost every one of them went broke; some of them several times. The reason was that the price of lumber was so low. The average return on a thousand feet of lumber, after you paid freight, was $15. In those days logs cost you more than $8 a thousand, delivered at the mill, up against the saw. You could probably saw at $2.50 or $3 per thousand, then you had piling and drying, that would cost another couple of dollars. Of course I had my tie business in the winter. What I made on the ties I lost in the mill. If I cut a million feet of logs and sawed them up and sold the lumber the following summer, if I had a thousand dollars clear at the end of the summer, I thought I had done wonderful.
—*Martin Caine, a pioneer northern lumberman who came to the Prince George area in 1918 and built the first planing mill there in 1939.*

During the last half of the war and immediately after it, a whole string of large, modern mills were built along the East Line, initiating a wave of forest-based industrial development which has continued, practically unabated, ever since. About 1915, a Chicago grain exchange operator named Frost built a mill at Willow River, and after running it for a few years moved it to Giscome. Frost's Willow River mill was the first of a rash of openings over the next couple of years. The United Grain Growers, a prairie farmers' co-operative which needed a lot of lumber to construct elevators and other grain handling facilities, built another large mill at Hutton. Roy Spurr built Penny Spruce Mills at Penny (it later became Red Mountain Lumber). Northern Lumber, a company which had provided wood for railway construction, took over Frost's mill at Giscome and called it Eagle Lake Sawmills, which ended up broke shortly after it opened. Determining who owned these mills seventy-five years ago is an almost impossible task today. The East Line was an incestuous place, with various people forming and dissolving partnerships, buying each other out, combining operations, moving mills from one location to another and renaming them. It is a pattern that has prevailed, in a somewhat restrained form, until the present time.

In 1923 the Winton Lumber Company, a family-owned firm founded in Wisconsin in 1889, bought the Giscome mill and called it Eagle Lake Spruce Mills. When the forests in the northeastern US had begun to disappear around the turn of the century, the Winton family had acquired cutting rights and mills in other parts of North America, including Canada. In 1906 they moved into Saskatchewan and, a few years later, obtained a mill at The Pas, Manitoba. With some misgivings they came to northern BC. They tried to run a conventional winter horse logging operation, and failed. They tried a railway show, and failed. After nine years they left, selling the business to Roy Spurr and Don McPhee. The latter had arrived in the country in 1925 and bought the newly rebuilt Sinclair Spruce Lumber Company at Sinclair Mills from its previous owner, a man named Fanshaw. The Wintons, however, would be back.

The East Line wiped out a lot of would-be timber barons in the early days. Within months of the war's end, eighteen mills with capacities of 15,000 to 150,000 feet a day were cutting lumber between Prince George and McBride. After all, the country was full of high-grade spruce. A transcontinental railway ran through the mill yards. A large drivable river, the Fraser, could be used to move logs cheaply. And to top it off, a second railway, the Pacific Great

Sinclair Spruce Mills on the Fraser River east of Prince George. The sawmill is at centre and the planer mill at right. BCARS 73609

In 1923, we were considering acquiring the Eagle Lake Spruce Mill at Giscome, British Columbia. This was the only time I remember C.J. [his father] and D.N. [his uncle] differing on a trade—D.N. wanted to buy and C.J. wanted to rent and decide whether to buy after a year. I advised not to go into Interior British Columbia at all because almost every mill there had gone through receivership and we knew little about the country.

We bought the Eagle Lake Spruce Mill and things went bad there with us from the very beginning. We had invested something over a million dollars and borrowed four hundred thousand dollars more. Our management was not good.

We tried to Minnesota sleigh log a mountainous country. British Columbia is a country that experiences "Chinook winds." In the midst of cold sleigh haul weather, with ice roads built to haul on, a warm wind would blow in from the Pacific Ocean and wash out the ice roads overnight. Also, the country was very hilly which meant runaways and wrecks. So we went railroading. We got logs, but too expensive logs to help us out when the Depression struck.

I, for one, after watching some nine years of failure at Giscome, saw no way out except to get rid of the property in British Columbia, pay our debts, take our licking, then devote our most intense efforts to saving The Pas Lumber Company by revamping the management.

—*David Winton, whose family built The Pas Lumber Company. The Pas returned to BC in 1954, with the purchase of the Hales-Ross Planing Mill in Prince George.*

Eastern, would reach Prince George any time (if the politicians could be believed).

But the East Line was no piece of cake. In the depths of winter it was often frozen, and sometimes bitterly cold for weeks at a stretch. Then it was just as likely to warm up, melting ice roads and bringing a horse logging show to a halt. In the summer it was either hot and dry, or wet and muddy. The ground was soft and difficult to work on unless it was frozen. Persuading workers to stay in the small settlements which grew up around the mills was always hard, especially in summer. Daily trains passed through, offering transportation to the flesh-pots of Prince George and Edmonton.

> First time I worked in a sawmill I was nine years old. I operated the tail end of the mill, handling the slabs and edges. Some of the slabs I just couldn't handle so I got help with them. I worked during Easter holiday and at the end of the holidays I was given a small cheque. We used to go to town once in a blue moon, once or twice a year—Prince George, which wasn't very big at that time. I remember I bought two bib overalls, two shirts, a pair of running shoes and a straw hat. I was the envy of the town! Those were pretty hard times. I was really happy, I was on cloud nine.
>
> —Neil McLean, whose family built and operated McLean Lumber Company at Shelley, east of Prince George, beginning in the 1920s. Neil obtained a degree in forestry from the University of BC and in 1961 founded S.N. McLean Forestry Services, which is now operated by his son.

The Wintons learned about all of this, as did many others. In time, those who could adapt learned how to log the East Line profitably, and the northern forest industry took off. One of these successful operators was Sinclair McLean, a Nova Scotian who had grown up farming and logging. In 1900 he migrated to California, and about twenty years later he decided he wanted to own a ranch in BC. As he was travelling north on the CNR, he got sidetracked and ended up contract falling at Hutton. The logging business looked good, and when that job ended he headed west to Fort Fraser where he got his hands on a bit of timber and a small sawmill and planer. When a fire burned his timber supply, McLean got on his horse and headed east, looking for more timber. He ended up at Shelley, where, with an investment of only $20,000, he bought timber and set up a 50,000-foot mill, under the name McLean Lumber Company, in 1923. A second mill, Shelley Sawmills, was built later. Under different names and setups, McLean operated there until 1951.

Hauling logs out of the woods on a slide-ass at McLean Lumber's logging operations near Shelley. Logs were loaded on a wooden frame which slid along greased rails. FFG 55

McLean was one of those rare individuals who, starting with very little and confronted with adverse conditions, devises creative solutions to apparently insurmountable problems and, in doing so, advances the entire industry. He found that the solution to maintaining a mill on the East Line lay in logging year-round, or at least for more than the winter months. How to do this was another question. Freezing weather was unreliable in winter and hard, dry ground conditions unpredictable in summer. Conventional horse logging techniques would not do. If he had had a lot of money, or been willing to sell control of his company, McLean could have gone into railway logging, but even then he would have had to find a way to skid logs to the railway economically and efficiently.

Instead he improvised, as countless loggers have done before him, by adapting methods used elsewhere and by inventing his own. The basic system he started with, to haul logs to the river, was a pole road. Poles were cut from easily available timber, then peeled and trimmed to 12-inch (30 cm) butts and 8-inch (20 cm) tops, about 40 feet (12 m) in length. They were laid 8 feet (2.4 m) apart on short blocks of wood, 18 to 24 inches (45–60 cm) long, spaced about 8 feet apart and stacked as high as required to maintain a steady grade. The poles were laid, butts downhill, with the top end of each pole mortised into the butt of the preceding one, with an overlap of two or three feet (1 m). A 3-inch (7.5 cm) hole was bored through both of them and a small sapling driven top-first through the hole and into the ground, then cut off flush with the pole. A three-man crew, with a team to skid poles, could build about 80 feet (24 m) of pole road a day.

Initially McLean used a device called a slide-ass to haul logs over the pole road. A set of bunks were built out of 10-by-16-inch Douglas fir timbers, about 11 feet (3.3 m) long. They were notched at each end to fit the curve of the pole rails and lined with forged steel plates. Two more timbers about 12 feet (3.5 m) long were bolted to the bunks, creating a rectangular frame onto which logs could be loaded and cinched down with chains. Large U-bolts through the cross timbers were attached to four- or six-horse teams, and the slide-asses were hauled along the greased pole road. There were no brakes on a slide-ass, so on steep downhill grades, hill brakes of the kind used throughout North America were employed. The slide-asses were used on hauls of 2 miles (3 km) or less.

Progress came in the form of pole cars. This device consisted of a frame, similar to a slide-ass frame, which was mounted on a set of concave cast-iron wheels. These wheels were

A Sinclair Spruce Mills slide-ass loaded with logs, ready to dump into the Fraser River, c. 1924. When the chains were released the logs were rolled off with peaveys. BCARS 73609

Top: A hill brake in operation with a slide-ass at McLean Lumber. Because slide-asses and pole cars had no brakes, longer grades of more than 5 percent required the use of this device. FFG 56

Above: Loading a slide-ass at McLean Lumber. Logs were skidded to the side of the pole road with horses and decked, then rolled onto the slide-ass with peaveys. FFG 48

Above: A pole car with concave cast-iron wheels hauling a load of Sinclair Spruce Mills' logs to the Fraser River east of Prince George, 1930s. BCARS 73618

Below: McLean Lumber's main fin boom on the Fraser River. A string of logs has been anchored in the river and is held in position by the plank fins on the left side of the logs. This boom guided logs along its right side for half a mile across the river, into the millpond. FFG 84

produced by most foundries making logging equipment, and they came in different sizes. Pole cars were hauled over the same type of pole road used with slide-asses, and could carry a bigger load for distances of up to 3 or 4 miles (5–6 km). Like slide-asses, they lacked brakes and were used with hill brakes.

The slide-ass or pole car was sometimes loaded by hand with peaveys, but usually it was positioned under an A-frame to be filled. A horse-drawn line ran through a pulley hung from the A-frame to a double or "crotch" line, each with a tong or end hook. The hooks were set into the log and when the horse pulled ahead, the log was raised into the air and guided onto the vehicle by two men using ropes attached to the tongs or end hooks.

Although McLean did not build his own railway, he did make use of the CNR (formerly Grand Trunk Pacific) main line which ran through his timber and past his mill. Pole roads were built to this line and logs 16 feet (5 m) long, or occasionally as long as 18 or 20 feet, were hauled with slide-asses and pole cars. Most of the logs, however, were floated down the Fraser to the mill.

The river driving system employed by McLean and a few others on the East Line may have been the first of its kind in BC. To move timber logged in the winter, a more conventional procedure was followed. Logs were hauled by sleigh or sloop to the river and decked along its banks. After the ice went out, they were rolled into the river and floated down to the mill. To assist in this process, and to facilitate constant log movement downstream during the summer logging season, a system of fin booms was used.

Fin booms were built with long strings of logs tied end to end with cable and anchored or tied to shore at strategic locations along the river. A fin—a 10-foot-long (3 m) two-by-ten plank—was laid on one side of each log in the string and the upstream end of it attached to the log. A piece of two-by-four 6 feet (2 m) long was used to brace the fin at an angle to the log. When the boom was placed in the river current, the fins pushed it over to one side of the river or the other. By adjusting the angles of the fins, crews could position the booms to deflect logs away from sandbars or rocks, and direct them into the millpond. During booming season a crew of young "river rats" constantly patrolled the booms, adjusting the fins to divert logs around emerging sandbars as river levels fluctuated. They also broke up jams, cleared logs off

A short fin boom, an adjustable set of planks and two-by-fours, used to direct logs around a sandbar on the Fraser River near Shelley. FFG 46

bars and rocks, and generally kept the logs floating down to the mill. When the system was properly maintained, logs could be dumped into the river at the logging camp, and they would drift downstream on their own, guided around obstructions and into the millpond.

Except for the occasional boom being swept away by a "sweeper"—a large spruce or cottonwood that fell into the river and drifted downstream, roots intact, knocking out booms and pilings as it went—the operation was largely a smooth one. But construction and maintenance of fin booms was a major task. They had to be removed each fall before the river froze, and replaced after breakup in the spring. A shorter boom could be anchored to a rock with an eyebolt drilled into it, or attached to a tree on shore. Larger booms, such as the half-mile-long main boom, were tied to heavy concrete piers built on the ice and sunk to the river bottom. Every year new pilings had to be driven. McLean had his own pile driver, a special float for moving boom anchors, and a fleet of riverboats for hauling supplies and tending the booms.

> The horses were all shod for hauling. The caulks were quite sharp. The toe caulk was an inch high and it came to a chiselled point for winter. And then the heel caulks—actually we never used the actual caulk. You could buy caulks and weld 'em onto the ends of your shoes but we used to just heat up the ends of the shoes and turn 'em up at right angles and then point 'em a little bit. In a falling, bucking and skidding operation they were pretty careful in swamping out their trails, because you can imagine if you didn't take care and you got your horses out there straining against a load and tripping over windfalls and brush, it wasn't that difficult for them to caulk themselves—run the caulk from their right foot into their left foot as they're stumbling along. They took quite a bit of care in making nice, neat trails to avoid that, 'cause once your horse caulked himself it would take quite a while for it to heal up. It was pretty serious. You could get infection in there. I remember a lot of the skinners, they used to pee in that wound. It was kind of a sterilant. They also had proper medication as well. The old blacksmith and farrier and barn boss, they were sort of jacks-of-all-trades. They had a real good knowledge of how to shoe a horse, how to apply veterinary skills and all that sort of stuff. They were pretty good rounded, knowledgeable men.
>
> In those days we cut the limbs right off nice and flush to the side of the log. Particularly if you're river driving, any log that wasn't properly limbed, when it hit that fin boom those little limb stubs would gouge into the side of the boom and the current would just flip it underneath the boom and it'd be a lost log. So they were pretty fussy about limbing those limbs right close to the log. Also it's much safer in the sawmill itself. When you're rolling logs with a cant hook, if you've got a log that's got a bunch of those knot stubs on it, it's just that much harder to roll.
> —*Neil McLean*

McLean's mill, like most of the others along the East Line through the 1920s and 1930s, was set up to supply the board market: it produced 1-inch boards from the high-grade spruce growing along the upper Fraser River. Most 2-inch planks were cut on the head saw, which was either a band or circular saw. The logs were turned to get the best grade possible: the mills were cutting for value, not volume. The rough planks were piled and air-dried, then run through a resaw that cut them into two 1-inch boards and planed them. The mill turned out

Above: Moving lumber around the McLean Lumber mill yard using a steel lumber wagon. Rollers with crank handles on the wagon allowed workers to load and unload stacks of lumber with ease. FFG 5

Right: Eagle Lake Sawmill's first crawler tractor towing a train of log sleighs near Giscome. Tractors replaced horses on longer hauls from the logging site to the Fraser River. BCARS 73528

Following pages: Eagle Lake Sawmill's single-axle Indiana logging truck, with the loading crew and driver, late 1930s. It is about to head off down a newly built plank road. BCARS 73525

1-inch boards from one-by-four to one-by-twelve in lengths of 12 to 20 feet (3–6 m), along with a variety of siding, shiplap, tongue-and-groove and other specialty items. Most of this lumber was shipped to prairie and US markets. In those days, before the Pacific Great Eastern Railway found its way into the north, East Line lumber was sold through brokers in Edmonton.

During the horse logging era on the East Line, which survived in diminished form until the 1960s, most of the horses were purchased from Alberta and came in unbroken, a stock-car load at a time. The first task was to break them to logging, pairing each new recruit with an older, experienced teammate. With luck a horse could work ten years; more often they were retired earlier than that with injuries. McLean, like most other loggers in the area, ran a farm along with his timber operation, raising feed for the horses as well as for the cookhouse.

In the 1930s East Line logging operations began to mechanize, bringing in trucks and crawler tractors. One of the first to do so on a large scale was Roy Spurr's Eagle Lake Sawmill at Giscome. His brother-in-law, Harold Mann, was a pioneer truck logger from the Coast who had grown up in his father's logging camps around Surrey. In 1935 he brought two single-axle Indiana trucks with single-axle Hayes-Anderson trailers up from New Westminster, a three-day drive at that time. It took another two days to drive them the 30 miles (48 km) from Prince George to Giscome.

> I learned to drive a logging truck on a plank road when I was thirteen years old. You got so that it was just the same as driving down the highway. You were used to it and that was it. If you fell off, you fell off. We were using frozen roads in the wintertime. Well, we had about three to four miles of plank road when we come off at Newlands, that's where we did most of our logging. You came off the frozen road onto the fore-and-aft timber road. We had a road crew. After the snow plough went through, the road crew would go down and they'd dig out every fifth crosstie on the road—that would give you a shadow when it was snowing, so you could see where the edge of the road was. If it built up and got icy we'd send a Cat down with what we called ice grousers. We changed the tracks on them in the wintertime to these ice grousers. They're not unlike a smooth pad that you see on a Cat today. We'd just walk the Cat down this plank road and it would crunch the ice all up.
>
> In the winter we ploughed the frozen road. Originally they had an old wooden D-plough on the old Indiana for ploughing. Just a wooden plough, nine feet wide. Then we got a little more modern and got a WD20 single-axle White with a dump box on it with a nine-foot plough, which was all hydraulic. That was my job for a couple of winters, ploughing. Lots of nights I'd leave camp at 12:30 or 1:00 to get the road opened up. The night foreman would wake myself and the guy that used to swamp with me and we'd go to the back end, open the main road and then go right straight through to the lake and get that all opened up. You could move snow with that machine. Once you got wheeling you could throw snow thirty or forty feet in the bush.
>
> To build a frozen road you'd start in the summer and build a rough grade with your dozers. Then as soon as it started freezing sometimes we'd have to work for frost. There might be twenty guys with snowshoes on that road, walking that road back and forth to work the frost in. If you travel something that's frozen you pound the frost down. Well, if it was too soft, we'd start that-away. Then when you got some frost in, you start maybe running the pickup over it and then, if nothing else, maybe the grease monkey or the mechanic, if they weren't busy. When you got some frost into the road you took the snow plough, once it would carry it, and drove back and forth on it with that. Just to work the frost in. We tried to keep the snow off but you'd be working during the day and it'd be snowing, so naturally you'd get some buildup. When our snowbanks got too high we'd just take one of the D7s down and he'd climb right up on the banks and push them back out. Myself or somebody else'd be on the snow plough and cleaning up after him so we had somewhere to get rid of the snow.
>
> —Bill Mann, a nephew of Roy Spurr. He worked at Eagle Lake Spruce Mill most of his working life.

At this time, in the midst of the Depression, the Giscome mill was one of the largest operating in the Interior and one of the earliest in the north, if not the entire Interior, to convert completely to trucks. The mill and logging camps ran year-round and Mann had to adapt his coastal trucking techniques to conditions along the upper Fraser. The crew trucked logs in the winter over frozen roads roughed out with bulldozers in the summer and fall. At the first snowfall and frost, twenty or thirty men were sent out on snowshoes to pack down the snow and work the frost into the ground. When the roadbed hardened a bit, a pickup truck ran back and forth on it, followed by one of the logging trucks with a snow plough mounted on it. Eventually the road got hard enough to support a load of logs, and then it was a matter of keeping it clear of snow. In those days it snowed 6 to 8 feet (2–2.5 m) a year in that country. Mann mounted a wooden V-plough on the front of a White gravel truck and assigned his son Bill the task of clearing the road every night after midnight. With the accelerator to the floor and no traffic to contend with, Bill could throw snow 40 feet off either side of the road.

In the summer, Eagle Lake hauled over fore-and-aft log roads and plank roads. A fore-and-aft road was built by roughing out a roadway and laying logs, two sets of three the width of the truck's wheelbase apart, on crossties laid every few feet. The tops were hewn flat to provide a running surface over which a careful driver wheeled a load of logs. They were tricky roads to drive on, particularly when it was wet or frosty. Occasionally they were used in

Above: A plank road under construction at Upper Fraser Mills, east of Prince George in the late 1930s. BCARS 73503

Above right: A three-man crew cutting planks for road construction with a portable mill, for Sinclair Spruce Mills, east of Prince George. BCARS 73622

Right: Loading Eagle Lake Sawmill's Hayes-Anderson truck. The loader consists of a winch mounted on a truck (at right) which hauls in cable, running through a block on the A-frame behind the logging truck, attached to the logs with tongs. The two loaders on the ground set the tongs, and the chaser on top of the load positions the log and releases the tongs. The fourth member of the crew operates the winch, while the driver relaxes. BCARS 73541

winter, after a snowfall. The road maintenance crew would uncover the end of every third or fourth crosstie so the driver could guess where the road lay. If snow or ice built up on the running surface, a crawler tractor with snow tracks was sent down the road to break up the packed snow and ice. If a truck slid off the road, it was usually no big deal; it was simply jacked back up onto the road.

Other roads, as well as the corners and turnouts of fore-and-aft roads, were surfaced with planks. Small bush mills were hauled out into the woods to cut 3-inch planks, 9 feet (3 m) long. These were laid on a double set of stringer logs set on crosslogs, and sometimes trestles. During the time they were used, from the 1930s into the 1950s, several hundred miles of plank roads were built along the East Line. Once an area was logged out, the planks were lifted and used on another road. Building a plank road was a labour-intensive task: a crew of three men needed a day to lay a hundred feet of framework alone, and the life expectancy of the road was about six years. Plank roads were maintained by the "road monkey," whose job was to jack up and block sagging sections, replace broken planks and attach guardrails to the ends of the planks along dangerous stretches.

The trucks used at the first truck logging shows were a mixed lot. Generally they were smaller than the logging trucks beginning to appear at coast operations by this time. Some, like the Indianas Mann brought north, were comparable to coastal logging trucks, but many were smaller. Mann, for instance, began with a Maple Leaf truck in 1939 and replaced it soon after with a three-ton Fargo. A truck was usually equipped with a single-axle two-ton trailer that typically carried a load several times its own weight.

At this time, before the advent of air brakes, the brakes on these rigs were primitive affairs. One of McLean's trucks, used mostly to haul lumber, was a Model A Ford equipped with a homemade trailer, built with the rear axle of an old International truck, that had a set of mechanical brakes on it. These brakes were operated with a piece of ¾-inch (2 cm) rope that ran under the trailer and through the rear window of the truck, and was tied to the dashboard. To brake, the driver slipped his right shoulder under the rope and heaved up on it at the same time he applied the truck brakes.

McLean Lumber's Model A Ford lumber truck with a trailer made from an old International truck. The trailer brakes were operated by a rope tied to the truck's dashboard.
FFG P996.3.11

> We used to have Cat houses. Not the kind that you can go and spend your money at for a woman. At nights when we shut down, the Cats would go into these Cat houses. They were a shiplap building on skids and there'd be a big barrel heater in there. In the night the grease monkey would go out at maybe ten or eleven at night to fuel and grease the machines, and when he finished servicing the machine he'd stuff the old barrel heater full of four-foot wood and go on to the next one. Kept everything warm so that they would be ready to go in the morning.
> —Bill Mann
>
> The first two seasons I worked on the river we couldn't afford tents. We used to just sleep out under the trees and when we woke up in the morning it was soaking wet in our sleeping bags. Then you jumped in the river, worked in the river all day. Seemed like you were wet all the time. Anyhow, we finally got a tent. They bought a secondhand tent from the Forest Service and we thought we were in seventh heaven, living pretty high on the hog there with a nice dry place to sleep at night.
> —Neil McLean

Loading the trucks required more machinery. Compared to other forms of logging equipment, a truck was an expensive machine: letting it sit idle while a couple of workers used peaveys to roll logs aboard, or a horse cross-hauled them, was not acceptable. McLean's solution was a "jitney," consisting of a 1930s Prince George cartage truck with a winch mounted on back, attached to a crotch line with two log hooks. The jitney towed an A-frame mounted on skids. The A-frame was positioned by the log deck and the jitney parked out of the way. The jitney operator paid out some line and the loading crew of two set the hooks in the ends of a log. The operator then reeled in the winch, raising the log over the trailer. Ropes on the hooks were controlled by the two loaders, who guided the log onto the truck with the help of the driver. With four workers involved in loading it was still a labour-intensive operation, but with practice they could do the job quickly.

A company that bought a truck usually also bought a crawler tractor, generally referred to as a Cat, short for Caterpillar, no matter what brand it was. The basic Cat had been around since the turn of the century. Until the late 1920s they were used by loggers in place of horses,

Right: Loading logs at a McLean Lumber skidway, 1930s. FFG P996.3.72

Below: A Model 30 Caterpillar crawler tractor skidding a sloop load of logs near Cranbrook, early 1930s. CFI

to skid logs and haul sleighs or sloops. In 1926 Bob LeTourneau, an eccentric California road-builder and machine designer who went on to develop a long line of unique electric-powered logging machines, invented the dozer blade. This simple innovation revolutionized truck logging across the continent: now there was a machine to build roads, and loggers could get away from the laborious process of laying fore-and-aft or plank roads.

When East Line loggers began buying Cats, they did so primarily to skid logs. Occasionally someone used a dozer blade to prepare a roadbed for a wooden road, but it churned up a lot of mud and the Cats were prone to breakdowns in this work. Further, Forest Service regulations in East Line forests stipulated that selective logging systems must be employed, to encourage the regeneration of spruce. Operating a Cat with a dozer blade could cause a lot of

> We logged all summer, and in October we had 12 million feet air-drying; and every stick of it was sold. It was October 1920, and Henry Meeker and I went down to Spokane to see the first crawler tractors. Jones, the superintendent for the Potlatch Lumber Company, had arranged to have a demonstration there. There were two tractors: the Hope, made in Stockton, California, and the Best, made in Peoria. Later they combined to produce the Caterpillar. There were about twenty-five operators there, and twenty-four of them said the tractors would never be worth anything, and they'd never stand up! At that time they had mechanics driving the machines and they were not accustomed to skidding logs. They ran right over stumps and twisted themselves around stumps. Later they trained men who had skidded with horses, and from then on the tractors were established.
> —Allen H. DeWolfe, born in Minnesota in 1887. He became a BC land surveyor in 1913. He designed logging flumes and logging railways throughout southern BC and in 1920 was superintendent of Henry Meeker's Nicola Valley Pine Mills at Merritt.

damage to the remaining stand, or to young growth. Consequently, during this period dozer blades were not permitted on Cats when they were used for skidding logs.

Eventually trucks and Cats solved many of the problems encountered in logging East Line forests, as they did elsewhere in the Interior. The Cat was not simply a mechanical alternative to human- and animal-powered technology: in this part of the country, mechanization in itself was not the answer. This was made quite clear by the Wintons' experience at Eagle Lake Spruce Mills at Giscome. The Winton people were highly experienced operators, with a lot of skill and solid financial backing. They were successful at most of their endeavours, before and after their disastrous experience at Giscome. They realized early that a conventional horse logging system would not work in the conditions found along the upper Fraser River, and they pursued the only other course open to them—mechanization. They mounted one of the most sophisticated and extensive logging operations in the Interior to that date. They had an elaborate railway network and two locomotives, a Shay and a Heisler, which they used from 1924 to 1931. On top of that, they purchased the most sophisticated cable

Skidding logs with a Caterpillar crawler tractor at an Eagle Lake Sawmill site near Giscome, 1930s.
BCARS 73533

> Typically there were some resident loggers that lived and worked right around the Prince George area and had little farms and so on, homesteads. But a lot of the logging fellows came from the prairies. You can imagine, you've got a farm in Alberta or Saskatchewan, you bring your harvest in in the fall and just about the time it starts to freeze up you've got spare time, you head to BC and go into a logging camp. You sign up for the winter to go logging and you stay there 'til breakup next March or April. Then you leave with your paycheques and head back to the farm on the prairie. So there was a lot of good loggers came from Alberta and Saskatchewan farms. There was quite a transient population of loggers.
> —Neil McLean

logging system available in those days, two Lidgerwood skidders.

Lidgerwood logging equipment was developed in the 1880s in New York by a Michigan logger, Horace Butters, to log cypress out of North Carolina swamps. The biggest Lidgerwood skidder consisted of a 270-foot (90 m) steel tower mounted on a heavy-duty railway car. It had an enormous set of steam-powered winches to operate a skyline, a main line and a haulback line to pull a carriage along the skyline. With an additional line, the carriage could be pulled into the machine independently of the main line, providing slack on the main line to facilitate the attachment of logs along a wide swath for a thousand feet or more from the railway line. The machine also had a large steel loading boom.

The skidder was mounted on an oversized rail car and pulled into place by a locomotive. The entire outfit could be raised with steam jacks and its trucks removed to enable empty log cars to pass beneath it. Empty cars were spotted on the rail behind the skidder.

> You had to pick the right location for anchoring a fin boom. We'd say, we've gotta have a boom in here someplace. So you start walking up and down the riverbank and you start throwing chunks of wood into the river and watching the current, to read the current, see how it goes. Then you'd pick your spot and say, okay, I think this is where we should put our deadmen or this is the tree we should anchor to, or the boulder, or whatever. It was quite an art to it.
>
> You couldn't just adjust the fin boom and leave it. First of all, the water's rising and falling and every stage of water affects the boom differently. So at one stage the boom would be too slack and the logs wouldn't miss that bunch of rocks in the river down below. They'd start piling into the rocks and causing big logjams. You'd see that coming and you'd have to go out there and adjust your fins to keep that boom over a little bit, so it's always in the right part of the current. Again, when it started rising up sometimes it would throw the boom too far, at too much of an angle to the current, and the logs would come and hit that boom and just flip underneath it. So you're always having to adjust those booms. Not only that, there was what we call sweepers. Trees'd come floating down the river, roots and all, and they'd hit those fin booms and sometimes they would shear off fins like nobody's business. You'd have to go and put 'em back on there because there's logs coming as well. And then of course there were those logs that in spite of your best effort would get hung up on the bars and on the rock centres, and you'd have to go with your peaveys and take 'em off.
>
> Our job was to patrol the river. There was the sawmill and then you'd have a couple of camps, river driving camps that were just tent sites, and then we'd have the logging camps up above. You'd be staying at one of those places and be patrolling back and forth, adjusting fin booms and rearing down logs, taking logs off the bars and breaking jams on the boulders out in the white water, that was fun. Get your adrenalin up.
>
> The logs came down the river and piled up against a big boulder. You might have fifteen logs, or fifty or a hundred logs piled up against one or two boulders out in this white water, really moving. It's just like a nail keg—they're all crisscrossed and it's one helluva snarlup against that boulder in that fast water. So you gotta go in there, nose the bow of the boat up the bottom end of the boulder and three or four of you get out there with your peaveys and try to find the key log. Usually if you find the key log and you break that one loose, it'll really get a bunch of logs moving. Boy, once that thing starts breaking, those big logs they start spitting out of there. You gotta get the hell out—I mean, talk about energy. All hell breaks loose! You can find the wings on your feet getting out of there and back into the boat. We used to have a great time running logs to get back.
>
> —Neil McLean

The haulback line pulled several chokers out into the felled and bucked timber where they were attached to logs, five or six at a time. At a signal, the main line was hauled in, dragging the logs to track-side where the loading boom was used to pile them on rail cars. An additional winch was used to move the loaded cars down the rail line, pulling an empty car underneath the skidder and into position for loading. For some thirty years the Lidgerwood skidder was the most sophisticated and expensive logging machine in the world, and it led the way to development of the elaborate cable logging systems which evolved in the coastal rain forests of the Pacific Northwest.

The Wintons bought their Lidgerwoods from Staples Lumber near Cranbrook. They were a bit smaller and lighter than the models used on the Coast, but enormous compared to anything ever seen along the East Line. A former Staples logging superintendent, an Irishman named Barney, came along to run the show. After a couple of years the company couldn't make the skidders pay, so it went back to horses and the skidders disappeared from BC logging history. It may have been because of a lack of workers experienced with cable logging systems, or because there was not enough big timber to make a cable logging operation economical. Or it may have been these and other factors, combined with the onset of the Great Depression in 1929, that forced them to sell.

The Depression hit most of the large Interior sawmills hard. Almost all the established operations in the southern half of the province which had not closed because of timber supply problems during the 1920s shut down because of financial problems in the 1930s. The upper Fraser River area was the only whole region to escape this fate, and even there, doing business in the 1930s was a dicey proposition for many. Sinclair McLean, who had made the mistake of selling all his lumber through one broker, was forced to close the Shelley mill for more than ten years. His second mill closed also, during 1933 and 1934. When things began to pick up a bit the following year, he set up some smaller mills, and eventually reopened the Shelley mill. Like many others, McLean survived the tough years by adapting to circumstances.

Some mills lost money consistently on lumber sales and made it back in other ways, such as by selling ties. During the early 1930s Sinclair Mills lost money on lumber and made it back on the cookhouse and the horses. The workers were paid $2–3 a day and charged a dollar or more for board. Teamsters were charged $2.75 a day to keep their horses in camp. Everybody struggled, and some of the mills managed to hang on until the late 1930s when markets began to improve.

Eagle Lake Sawmill's Lidgerwood steam skidder working near Giscome, late 1920s. The steel tower (at left) supports a skyline along which the carriage (top right) runs. The mainline runs through the block on the carriage and is attached to the log, suspended at right. The loading boom (centre) loads logs onto the rail cars. This machine first worked in BC at Staples Lumber.
BCARS 73542

I can remember one time when I was about ten years of age. My dad was just nicely getting established, still under construction. He was having a real hard time. He had borrowed money from the bank to get it going, there was no production, the money was all out to build the mill. These three guys come up to him and said they weren't going to skid any more logs 'til he paid 'em more, and he said, Well, I can't pay you more. I've paid all I can. Until we start shipping some lumber it's just out of the question. So they said they weren't gonna do a damn thing. And one guy said he was going to get money out of him one way or the other. He come over and he was gonna take a swing at my dad, and my dad gave him a hard left and a right to the jaw and settled the labour negotiation and they all went back to work. I was scared fartless.

I saw that again in the '40s. I remember one morning my dad settled the same kind of thing. A big planer man, he was a real big man, six-foot-six and well over two hundred pounds. He was sort of the bull leading this thing, and he was calling a strike. My dad went up to him and said, There's not gonna be any strike today, you're going back to work. And give him a couple of hard good ones to the jaw and they all went back to work. That was labour negotiations in those days.
—Neil McLean

CHAPTER 6

Hacking & Hauling

The 1930s

While the mills along the East Line were being built during the 1920s, a very different kind of forest industry was developing along the railway west of Prince George. After crossing the Fraser River, the railway followed the Nechako River to Fraser Lake, then up the Endako River to the height of land and down the Bulkley River to the Skeena. The forest cover in this last stretch, down the Bulkley Valley, was much different than along the Fraser. It consisted mostly of lodgepole pine, for which a market had not yet developed.

Before the coming of the railway, limited development of the forest industry had taken place in this country—only enough to serve the local market. One of the mills, a water-powered sawmill built near Endako in 1910 by Charlie and Howard Foote, was unique because the only metal used in its construction was a whipsaw blade. The dam, water wheel, millworks and carriage were all built of wood and fastened together with wooden pins. The carriage consisted of two spruce timbers with V-shaped grooves cut along the lower face, mounted on two more beams with inverted V-shaped upper faces. The timbers were greased with bacon fat, as were the rest of the mill's moving parts. This simple but effective setup functioned for several years before being destroyed in a forest fire.

In the winter of 1912, John Ruttan sledded a boiler and steam engine into Fort Fraser and had it operating by the following summer, cutting lumber for the town growing there. When Sinclair McLean arrived from California a few years later, in 1919, he purchased this mill and operated it for a while before relocating it on the East Line.

Construction of the Grand Trunk Pacific line between Prince Rupert and Prince George began on the coast in 1909, and most of the construction materials were cut at the Coast and shipped east along the railway as it was built. This portion of the line contributed little to the development of the forest industry in the area, with one notable exception: ties, which were provided by one contractor, Olof "Tie" Hanson.

Hanson was a Swede who immigrated to the USA at age twenty and eventually homesteaded east of Edmonton, along the GTP line, which was then under construction and moving west. To finance his farm, he worked winters cutting ties in railway construction camps. A reliable, honest worker, Hanson soon obtained subcontracts for tie cutting. He found the work more to his liking than farming. In early 1907 he rode to the end of the line, near the BC border, and spent nine months walking across the province to Hazelton, examining the timber along the railway's proposed 800-mile (1280 km) traverse of the province. By 1909 he had set up the Hanson Lumber and Timber Company in Prince Rupert and obtained a tie-cutting contract for construction between the Coast and Endako.

Hanson hired close to 1,000 broadaxemen from Sweden and eastern Canada, then set up twenty to thirty camps in the area between Terrace and Usk, and hewed tens of thousands of ties from the hemlock stands there. On completion of the railway in 1914, Hanson turned to pole making, cutting cedar along the Skeena River and floating poles to loading points on the railway at Cedarvale and Skeena Crossing.

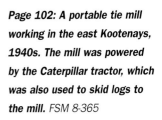

Page 102: *A portable tie mill working in the east Kootenays, 1940s. The mill was powered by the Caterpillar tractor, which was also used to skid logs to the mill.* FSM 8-365

Page 103: *Hauling hand-hewn ties in the Bulkley Valley, 1920s.* BVM P1145

A 1930s tie camp near Burns Lake. Strimbold collection

At the end of World War I, a huge market for hewn pine ties developed, and Hanson entered into it enthusiastically. By this time hemlock ties had been found to be susceptible to rot, while axe-hewn pine was more resistant. Because little or no lodgepole pine lumber was marketable until the early 1960s, about the only commercial use that could be found for it was railway ties.

The growth of the tie market coincided with the arrival of hundreds of settlers in the Bulkley Valley. Lured to the area by free land, the newcomers found their new homesteads covered with dense stands of lodgepole pine, which they had to clear to earn title to the land and to grow crops. Many of the settlers were Scandinavians and experienced woodsmen. They needed cash, and they gravitated quickly to the tie camps Hanson established along the upper reaches of the Bulkley and its tributaries.

Within a couple of years, Hanson and another contractor who had set up shop in the area had evolved a better organization of the tie business. Sivert "Bull River Slim" Anderson arrived at Burns Lake in 1922 after spending a couple of years working in CPR tie camps on the Bull River. He and three partners obtained a timber lease and, during their first winter in the northwest, cut 27,000 ties. After another successful winter, they set up a permanent operation at Decker Lake. That same year Hanson moved his headquarters to Smithers. He dominated the business from Hazelton to Burns Lake, while Anderson took over from Burns Lake to Endako.

I remember one place I was packing ties after I quit cutting logs. Man, that's the most hardest work I ever did. They cut up to seven hundred ties in a day. Just one guy. Just me was piling. You run, you don't walk. You pile it higher and higher. By the time you run back there'd be another tie coming off. And sometime with a big jack pine, two number ones or two number two ties will come out of it. Most hardest work I ever done in my life. Oh man.

—*John Dokkie, who started logging and tie hacking at Moberley Lake in the 1930s.*

Viola's dad, he was in the tie business. He loaded ties. Talk about a tough old Swede now. He packed ties on his shoulder for twenty-five years. Up to three hundred pounds each. He talked about at Fraser Lake, there was four thousand ties froze in the ice and they got an order for them—in the '30s. So him and Pete Olsen went down there and chopped 'em out of the ice, picked 'em up and put 'em on their shoulders and packed 'em up the jack ladders, put 'em in the boxcar.

"Yooo," he said, "that ice, she was tough on the neck!" But imagine what those would weigh.

—*Archie Strimbold*

We have another bit of trespass on our hands, Willow River Lumber Co. (Etter & McDougall), involving perhaps half a million. Report with all details has gone forward. Jack McDougall is business head of the firm. Jack McDougall's father, no longer actively interested in business, is a conservative of the old school. He makes no bones about telling any Civil Servant that he has his job through pull, that it will only last as long as the government lasts, and to make the most of it while he may. He is also firmly convinced that all the evils that business is heir to are the fault of a rotten Liberal Government too long in power. How much of that Jack has imbibed I have no idea. He never makes any reference to any such topics. I have the impression though that it would not hurt him a bit to put something over, not nearly so much for the value of the timber he got, as the kick he would get out of doing it. I think this trespass is deliberate but we have nothing whatever to indicate the fact and judging from what we can put on file it is unintentional and I have made my recommendations on that basis.

—*C.D. Orchard, then Prince George district forester, in a letter to Chief Forester P.Z. Caverhill, March 13, 1928.*

Above: Archie Strimbold, as a young boy, working his family's farm near Burns Lake. Logging and tie hacking along the rail line west of Prince George were winter occupations; for the rest of the year the men and horses turned to farming.
Strimbold collection

Right: The 150,000 hand-hewn ties in this deck, pictured in 1919, were floated down Shovel Creek, east of Burns Lake, and flumed to a railway siding. BVM P2425

Most ties were cut within a few miles of the railway, but some operations went much farther afield. Jack Stanyer and his sons cut ties at the west end of François Lake and towed them down the lake with a steam-powered sidewheeler he built in 1923. The ties were driven down the Stellako River to Fraser Lake. There they were caught in a boom, towed down the lake and loaded onto rail cars with a jack ladder.

The winter production of logs and ties at Weatherhead's camp near Yahk, 1925.
BCARS 56360

As the large stands of timber capable of supporting big tie-cutting camps disappeared, Anderson and Hanson changed their methods. They obtained tie orders from the railway and subcontracted them to individual tie hackers, most of whom were local settlers. They cut ties off their land, or other land, in the winter and proved up their homesteads the rest of the year. This system functioned smoothly for almost a decade, producing millions of ties, developing farms along the Bulkley and building the foundations of a permanent forest industry. By the end of World War II, many settlers had revised their expectations. Instead of devoting themselves to the clearing and farming of marginal agricultural land, they came to view their homesteads as places to live, grow food, raise enough livestock for their own use and provide a base for their lives as loggers, millworkers or sawmill operators. The economy which developed in the valley was more diverse than the agricultural community that the settlers had first envisioned.

Making or "hacking" ties required a great deal of skill with an axe, as well as physical strength and endurance. The ideal trees for ties were about 12–16 inches (30–40 cm) in diameter, and lodgepole pine stands in the Bulkley Valley were well stocked with such trees. Most tie hackers worked alone, in camps or on their own farms. The task was fairly straightforward: select a suitable tree, fell it with a one-man crosscut saw, measure it into as many 8-foot (2.5 m) lengths as possible and delimb it with an axe. The top was left on, along with its limbs, to steady the tree during hewing.

Standing on the tree, the hewer scored both sides of it with a double-bitted axe. To do this he made a series of cuts of precisely equal depth, about a foot apart. This was the most difficult part of the operation: if the scoring cuts were not the proper depth, it was impossible to cut a smooth surface during the next phase.

Once the tree was scored, the hewer used a broadaxe sharpened on one side only to flatten the tree on two sides, leaving a uniform 6- or 8-inch (15–20 cm) thick log, depending on the size required. He stood on the log to do this, using a hacking stroke rather than a swing, to leave a smooth, flat surface. Then the log was bucked into 8-foot (2.5 m) lengths.

Part of the tie hack's job was to drag or carry the finished tie to a skid road and pile it. Since each green, frozen tie weighed 200 pounds (90 kg) or more, the ties were usually dragged over the snow with a picaroon, an axe-like tool with a hook instead of a cutting blade. For this work, a tie hack received about 20 cents. Good hacks produced and stacked forty to fifty ties a day, and average workers might turn out twenty to thirty. A few exceptional individuals could cut a hundred ties in a day.

Tie hauling was organized in several ways. In the camps, the work was usually contracted. A farmer either paid a neighbour to haul his ties, or used his own team. A worker with his own team worked part of the day, cutting thirty or forty ties—about the load limit of a farm team—and stacked his load on a sleigh. In the Bulkley Valley, ties were usually stacked lengthwise if a single team was used, and crosswise on larger sleighs with four-horse teams. The practical hauling distance was no more than 5 miles (8 km). Most ties along the Bulkley River were hauled directly to the railway and stacked along the tracks. In a couple of cases, the Hanson company drove ties down creeks and used flumes to move them out of the water onto piles along the railway.

When the railway came to collect the ties, a loading crew consisting of four loaders and an inspector worked its way down the line from siding to siding, counting, grading and loading each hack's ties onto flatcars or into boxcars. Loading was a simple, if demanding, task. The loader upended the tie, stooped under it and stood, balancing it on his shoulder. He then ran with his burden, to gather momentum to make it up a 16-inch-wide (40 cm) plank leading to the rail car. After piling his tie in the car, he repeated the operation—four hundred times a day. At 3½ to 4 cents a tie, loading was the highest paid work in the industry. Standard hourly wages ran between 35 and 40 cents in those days, and a team of horses complete with teamster could be hired for $7.50 a day; good loaders made $15 to $20 a day. As they moved along the line, they stayed with the tie hackers' families, paying generously for their room and board.

The inspector was an important figure in this business. Each worker had his own timber mark and the inspector had to tally each hack's production, as well as ensuring buyers' specifications were met. He rejected anything he did not find suitable. In the Bulkley Valley, inspectors were known as "God's brothers," and their judgement calls were of great significance to the cash-strapped stump farmers.

From the time I was a kid the changes have been quite a bit. We started falling like with our crosscuts and horse logging, mostly ties and poles. My dad before me, he was a tie hacker, and all his brothers who lived in the valley here too, they were tie hackers and tie loaders, for Hanson Lumber when they had their thing, and afterwards it was Gordon Jewel took over. I worked for Gordon Jewel. Up at the other end, up at Houston, was Harry Hagman, the Finnish fella that had a big mill there. When we were young, it was a tie market more than a lumber market. When I was fifteen, I was loading ties and hauling 'em out with a team of horses from back of Quick Lake. It was tie hackers in the bush. Some sawmills sawing 'em, but mostly just hackers. You go out on a real cold day in the wintertime, you could hear them. It sounded like a moose out in the bush. There'd be maybe ten or twelve hackers out there and you hear this guy go "whew" and an axe'd come down. They blew their air out, a lot of 'em, eh, when they'd swing those big broadaxes. You could hear it running across through the bush.
—Bruce Kerr, whose family settled near Telkwa in 1918, has logged and farmed in the area all his life.

Tie hacking underwent a major change when a creosoting plant was built at Edmonton in 1927. Now the unhewn faces of ties had to be peeled, a task abhorred by most tie hacks. As a result, children growing up in the valley during the late 1920s and 1930s were put to work peeling ties for 3 or 4 cents each. The next step was to require that ties be hewn to a 10-inch (25 cm) width if they were to be creosoted. The result of these new requirements was to encourage the production of sawn ties. Before creosoting technology was introduced, most railways preferred hewn ties because the smooth faces were more rot resistant. But sawn ties were just as acceptable for creosoting as hewn, and treated ties lasted three times as long as untreated ones. As experienced hacks got older, they were gradually replaced by sawmills or hackers who had portable mills, although a few operators were still producing hewn ties as late as the 1950s.

A Nicola Valley tie mill, 1940s. It was powered by a farm tractor. NVMA P-L357

During the 1920s, along with the tie-hacking industry which developed adjacent to the Grand Trunk Pacific line, a relatively large stationary sawmill evolved at Fraser Lake. In 1919 Dan Webster and Bert Black, steam engineers from Eagle Lake Sawmills at Giscome, built a steam-powered mill. Over the next fifteen years, always short of cash, the two entered into a number of partnerships and gradually built up the mill to produce a wide range of lumber, lath, mouldings, windows and doors for the local market. They logged on François Lake and, like Jack Stanyer, built a wood-burning steamboat, a stern-wheeler called the *Bluenose*, to tow logs to the outlet into the Stellako River. An annual log drive moved the logs down to Fraser Lake and into a boom that was towed across to the mill.

◄

A crew of CNR inspectors and tie loaders filling boxcars at Bednesti, west of Prince George, 1925. The photograph was taken by J. Miles Gibson, district forester at Prince George and, later, dean of forestry at the University of New Brunswick. PGL 1637/1

Black sold out to Webster in the early 1930s, and when Webster died in 1933 he left his share of the mill to his girlfriend. She sold out to Mark Connelly, who guided the company through the Depression with such innovative marketing activities as selling eighteen rail cars of aspen logs to a match factory in China in 1934. The operation survived and eventually became part of West Fraser Timber Company.

In 1928 the beginnings of another major mill arrived when a young immigrant Finn named Harry Hagman crawled off a railway flatcar in Houston, alone and broke. He had lost his stake in a poker game on board ship, disembarked in Montreal and ridden the rods across Canada. In Houston Olof Hanson hired him to work as foreman on pole drives down the Skeena River. Two of Hagman's brothers joined him, and they worked through the Depression logging, hacking ties and making poles for Hanson, who also loaned them the money to build their first sawmill during World War II.

The mill was a fairly primitive affair, with a gas-driven circular saw and a carriage pushed by hand. This was the modest start of Buck River Lumber, which later formed Bulkley Valley Pulp & Timber and ultimately became one of the major components of Northwood Pulp & Timber Ltd.

Elsewhere in the BC Interior, the Depression was hard on the sawmill industry. Dozens of small operations and at least three major mills went under. Nicola Valley Pine Mills, which had been taken over by Henry Meeker in 1909, had become thoroughly integrated into the Merritt economy. In 1928 the mill began supplying the city with electricity. As the initial effects of the Depression sank in, local citizens agreed in a referendum to guarantee a bond issue put out by the mill—even though their mayor opposed the transaction. Ghost towns were appearing

I went to work for the Altmonson lumber company down in Parson in 1937 and they were logging then in the north fork of the Spillimacheen. It was fourteen miles to their camp in the north fork. Freight was hauled with a team to supply the horses in the camp and the groceries, because they had around fourteen men in the camp. This would be a couple of saw gangs, each two men and an undercutter, as was quite common. They had two gangs for falling and then they had skidding horses, a skidding team, usually a skidding horse. And they had the freight team, which they kept in the camp to skid when they didn't need supplies. And they had sloop teams. If they got a little too far from the river, a thousand feet away from the river, then they put them on a set of runners with the tail end of the logs on the ground—and these logs were all bucked to sixteen feet in the bush—and this sloop load was taken to the river and dumped and rolled into the river. Rollways all along the river. They piled them up on the ice. All they left was a little bit on the far side of the river for the river water to pass in the spring. Throughout the winter they usually put up around a million feet that way with fourteen men. By today's standards that was pretty slow logging.

Those fellows working in the camp, they went in there in the fall and they stayed there all winter. They wouldn't come out for Christmas, a lot of them. Single fellows a lot of them were. Swedish and Norwegian and European. They'd keep all their cheques 'til spring. They'd come to town with six cheques in the spring and they would cash them. They were always good. But in about three weeks they'd be back in camp sick and bedraggled and broke. Ready to start work for the summer, around the mill and the yard.

The logs were put in the river and in July, when the hot weather came, they sent a crew from the mill and the planer to bring them down the river, around by Spillimacheen, down to Parson. And that kept them sawing all summer until freeze-up. Being a steam mill they couldn't run after freeze-up. Middle of October was the end of the milling. They would plane for a while but it was steam also. All the lumber that was sawn that year was put in piles in the yard and kept 'til the next year. It was kept and planed the following spring. This went on until war broke out and then the demand for lumber was so keen that they started selling what they called "hot lumber"—right off the saw, through the planer, into the boxcar and out to Calgary. Green lumber. I just assume that some of those buildings had some pretty wide cracks in them.

You went along the bank and climbed on the jams, some of them a thousand logs. Of course that would back the water up. The idea was to find where the key log was. You went down in front of the jam to break those logs off the rock. You kept on doing this until the whole thing went. You made a few dashes for shore when things started to move. When they didn't, you went back and tried again. Finally, finally, they would go. And with all this water pressure behind they didn't lose any time, I'll tell you. Sometimes you'd assess the situation, sometimes you ran for shore across the front of it. Other times you got on top and got across the back as quick as you could and into the water, because the water was deep there and not moving and you'd get to shore before it started to move.

The biggest problem was you used to wear caulk boots, Paris boots. The leather would soften and it would dry and the caulks would go through the boots and cripple you. I've got corns today from that.

The outfit I was on, there was only one man ever lost and that was opposite Giant Mine. I wasn't on the crew when he was lost. But Columbia River Lumber before them, they lost a whole boatload of men. They lost ten men in one accident. The boat upset and in that white water they couldn't get out.

I did have a narrow squeak. Another young fellow and I were in rough water. We had a jam on a rainy morning, which didn't make much difference because you were wet anyway. His caulks were gone. We'd been there a couple of weeks and his feet were in bad shape. So he put a pair of rubbers on, which is a terrible mistake, to walk on a slippery log. Anyway we worked on a jam and it started to go. We both were gonna run to shore, and he slipped in. I wasn't thinking at all. I jumped right in after him and got ahold of him. There were logs banging into us and pushing us. I thought to myself, "This is it. I'm not gonna get out of here." We were being carried along. I don't know how we stayed on top even. We swept through a huge clump of willows that had broken off into the river but were still rooted. I hung on. I was still hanging onto him and he got a hold of those willows, and somehow we got out of there. We were sure banged and bruised when we got out. The old Norwegian foreman, as soon as he found out we could still walk, he says, "Well, we is bound to get on with it."

—*Orond Brashier, who has logged and operated sawmills around Parson for most of his life.*

throughout BC at this time and the people of Merritt hoped to avoid the same fate. In 1933 the mill went broke, dragging the city down with it. Merritt was placed in receivership, where it remained for twenty-five years. Meeker moved to Mission to found a shake and shingle empire, and the Merritt mill was picked up by Hugh Leir from Penticton.

Part of a winter's cut of ties at a BC Spruce Mills camp near Lumberton, ready to flume to a railway siding.

Following pages:
A team of horses is used to parbuckle logs into the BC Spruce Mills flume near Lumberton, late 1920s.
BCARS 73576

> My first job was sleigh hauling. I quit school just past Grade 8, when I was fourteen, and started sleigh hauling off the mountain into the river out here, down the Glenmary. My dad was sick. He got hurt in the '30s and he couldn't work, so I took over the horses and started sleigh hauling for George Bucknell up on the mountain. He had a small logging show. He always done all the falling and bucking and I done the skidding and sleigh hauling to the river. I stayed there until 1940 and then I started working for Tom Malpass and I started driving logging truck. I drove just about four years for him before I went on my own. I never had brains enough to quit.
> —Audrey Baird, who was born in the Shuswap valley in 1923. He was a pioneer Okanagan truck logger.

In the Kootenays, two big mills were lost. J.B. Winlaw's 75,000-foot-a-day mill at Wynndel, which had struggled through most of the lean years, was destroyed by fire in 1937 and never reopened. BC Spruce Mills closed for somewhat different reasons. After World War I, when it was taken over by Wisconsin interests that invested $2 million in it, the Lumberton mill cut 165,000 feet a day and was by far the largest in the Interior. Its timber supply was obtained from along the Moyie River and was moved to the mill along a 19-mile (30 km) flume, the longest ever built in BC. The mill survived until 1938, when the big logs for which it was built ran out. The only choice available to the company was to convert to a truck-based logging operation, which was not financially attractive at the time. Logging ceased and the sawmill gradually wound down as it cut through its accumulated inventory. It shut down entirely in 1940, and within a few years the thriving town of Lumberton, population 250, had disappeared entirely.

Those mills which did survive did so through the ingenuity of their owners and often the trust and support of their employees. When the Depression struck, Waldie's mill at Castlegar was caught with a lumber inventory of 9 million feet. Keeping the mill operating with this much lumber waiting to be sold was difficult. When orders could not be filled from inventory, workers were called in for a day or two. Often there was no cash to pay wages, so workers were given scrip, slips of paper indicating the earnings they were owed which were accepted by local general stores. When wholesalers delivered supplies, their deliverymen collected Waldie's slips of paper and exchanged them for lumber at the mill. This system worked both ways: once Waldie shipped a load of lumber to Alberta and received a shipment of beef in return. The beef was cut up and given to the workers in lieu of cash.

The spirit of helping each other out extended beyond the mill. Throughout the Depression the boiler room at the Waldie mill was available to transients as a place to sleep for the night. The cookhouse supplied breakfast in the morning, and packed lunches when the men departed.

On one occasion, when the cook refused to make lunches for the mill's "guests," Bill Waldie fired the cook and made the lunches himself.

> 1930 was a good year for us. Mind you, when we got into '31 we couldn't sell. We couldn't find buyers and we shut down. Not permanently, but we would shut down and run, start and stop, start and go—maybe three days a week and then two days a week, just sawing up logs we had in the water. No logging was done for two years, '30 and '31. In '33, Henry Gopp, a local fellow, had some timber up here that he wanted to put in the water for us, so we made a deal with Henry and got about a million feet of logs. This got us going and everything kind of broke for us. We were able to sell the lumber without losing too much money, and then we gradually expanded out again a little more.
>
> Prices then were very low. Mind you, we were getting the stuff pretty low and got help pretty reasonable too. We had losses in the early years, the '31 to '32 period, when we were repricing inventory. We had losses then all right, but these new logs, they came in and we got them for a good price, manufactured them for a low cost, got the costs down.
>
> —*Bill Waldie. He began working in his father's mill, Edgewood Lumber, in 1920, and managed it until it was sold to Celgar, and closed in 1961.*

With the outbreak of war in 1939, the forest industry was thrown into high gear almost immediately. Before the war, the largest importer of lumber in the world had been Britain, which obtained almost all its timber supplies from Baltic countries. With the Baltic closed to trading, Britain turned to Canada for timber, and most of that need was filled by BC lumber mills.

Rolling logs into the BC Spruce Mills flume at Lumberton. The logs have been skidded to the flume and decked on a skidway with horses.
FSM F.S. 8 262 (252)

I was too young for most of the log drives. I did some of my own when I had my own timber sales later on and I did take one pole drive down. I started out originally with my own timber sale before I was eighteen. You could do that in those days. My partner and I, he got one and I got one, and we went out with horses and crosscut saws and started logging. Those days are all gone now. Things turned out reasonably well. Stumpage was 35 cents a thousand for fir. You had to work pretty damn hard to make $2 or $3 a day. My partner was Charlie Hawes, he lives not that far away now. We were partners for a number of years. No official partnership or anything, we just worked together, did things together. He was older than I was, a few years. We started out cutting cordwood, firewood, in the '30s, 1936 or '37. We got, I think it was, around $1.25 or $1.30 a cord on the side of the road for a truck to pick up. And we paid half of that to the land owner. The guy that bought it was a dealer here in Vernon. If we worked like hell we could get a cord a day each. Two cords a day. It was all birch. Nobody wanted anything else. That's how we got started. Things didn't cost very much, but it was tough in those days. I still can remember when we first started out I couldn't afford a handle for my crosscut saw, so I had to wire a piece of wood to the end of the saw so I could use it.

—Len Bawtree, who was born in the Shuswap Valley and spent most of his life working in the woods there. He was logging superintendent for Riverside Forest Products and retired to manage his Woodlot Licence and private forest land. He also served as MLA for the area.

A junction of three flumes.
BCARS 73480

The demand was enormous. For more than two years Canada was Britain's only ally capable of providing substantial material and military assistance. The Canadian government established a Department of Munitions and Supply under C.D. Howe, who recruited leading industrial figures to mobilize for war. To organize the forest industries, he called in H.R. MacMillan and appointed him Timber Controller. MacMillan was given the task of organizing the entire Canadian forest industry to provide timber products.

One of the first Canadian undertakings was to build a network of almost one hundred airports where Commonwealth air crews could be trained. Lumber was needed for hangars, shops, instructional quarters, barracks and a thousand other items. Wood was considered a strategic material: in many applications it could replace steel, which was desperately needed to build ships, tanks, machinery and weapons. By the spring of 1940 every sawmill in BC was working flat out. When a shipping shortage developed that impeded the transportation of BC lumber to Britain, MacMillan arranged for its priority transfer to Atlantic ports via the transcontinental lines of CNR and CPR—an agreement that was highly advantageous to Interior mills located on the main railway lines.

The mills' accumulated lumber inventories were cleared out almost overnight. Conventional practice in the Interior at this time was to air-dry lumber, letting it sit for weeks or months to lose its moisture content naturally. This method had to be abandoned. Now lumber was cut, planed and shipped as fast as logs could be brought to the mill. It dried in place, on the walls of buildings hastily being thrown together across the country. In 1942 logging and sawmilling were classified as essential industries.

The shortage of strategic materials made it difficult, if not impossible, to obtain new sawmill equipment to help meet this demand. Nevertheless, a whole string of new mills appeared throughout the Interior. In the far northeastern corner of BC, where no mills of any kind operated, the massive Alaska highway project got underway. Slated to run from Dawson Creek to Fairbanks, it was 1,500 miles (2400 km) long and needed 130 major bridges, almost all of them to be built of wood. A lot of small, portable mills were moved in, some of them owned by individual contractors. One of these was Dan Pomeroy in Fort Nelson. After the war he sold his mill to Charles Bumstead, who barged lumber down the Fort Nelson, Liard and Mackenzie rivers to Inuvik, as well as trucking it north to Whitehorse.

The BC Spruce Mills flume at Lumberton, winding through a logged-off area, 1920s. Loading sites can be seen along the skid road. BCARS 73603

The most successful of the highway sawmillers was Gordon Moore, a local entrepreneur who operated several mills during road construction and, after it was finished, acquired several more which he operated under the name Fort St. John Lumber. Moore was a pioneer of the forest industry in the Peace River country and one of the originators of the bush mill system of lumber production which soon appeared throughout the Interior. Small three-man sawmills were hauled into the woods, where they cut logs available in the immediate area before moving on to the next stand. Their rough lumber output was trucked to central planer mills and finished. Moore set up his first planer mill at Dawson Creek and built up an extensive business which eventually became part of Canadian Forest Products (Canfor).

> We have had a quiet winter. All the operators have felt the effects of the market failure of 1929 and logging programmes have been cut to the limit except in the case of Eagle Lake Spruce and I understand now that after a full winter's logging they do not intend to do any summer logging. All the others have worked it the other way around. Have pretty nearly cut out their winter logging with the intention of making up in the summer should the market show any marked revival. In any event they all have much larger yard stocks than heretofore and several of them, Eagle Lake, Sinclair Spruce, Longworth, etc. could fill orders for pretty nearly a year without sawing at all. We have had pretty nearly as many operations as usual but all small so that we have been just comfortably busy without many complications.
> —C.D. Orchard, then Prince George district forester, in a letter to Chief Forester P.Z. Caverhill, February 20, 1930.

When the war began, Dave Martens was renting a farm in Saskatchewan and trying to scratch out a living there. After years of fighting drought and economic depression, he packed up and moved with his family to a homestead near Vanderhoof in 1942. He brought along some horses and went to work logging with his three sons. After a couple of years his oldest son, Bill, bought a $500 sawmill and powered it with an old Chrysler car engine he had picked up for $25. They logged all winter and spent the summer cutting the logs into rough lumber which they sold to a planer mill in Prince George. This operation, along with a second mill they started, Blue Mountain Sawmills, evolved into Plateau Mills, which today is a division of Slocan Forest Products.

At Kamloops, in 1943, Phos Bessette set up a mill called BC Interior Sawmills to feed the wartime market. It too grew to become a substantial operation after the war. That same year, in Merritt, the hapless Nicola Valley Pine Mills changed hands once again. Ken Long bought the mill from Hugh Leir, along with an understanding that Penticton Sawmills would not acquire timber rights in the Nicola Valley. At the end of the war, Long sold the mill to Allan Nicholson, a partner in Cranbrook Sash and Door when it was founded in 1904. After buying his way in and out of several mills in various parts of the province, Nicholson succeeded MacMillan as Timber Controller and, when the war was over, got back into the lumber business. He changed the mill's name to Nicola Valley Sawmills and kept Long on as manager. Long eventually bought the mill back in 1961.

The difficulty of obtaining new mill machinery during the war resulted in the recycling of old mills, and a good portion of wartime lumber came from mills which had closed during the Depression. In 1942 the Smith brothers, who had built the Armstrong Sawmill just after the turn of the century and added another mill at Salmon Arm in the 1920s, took over the old Rogers mill in Enderby. It had sat empty since its sudden closure in 1921, and most of its equipment was still intact. The Smiths started it up to feed the wartime market and continued to operate it until 1970 when it was sold to Crown Zellerbach and closed.

Along the Columbia River, an old-time sawmiller named Jim Parkin scratched out a Depression-era living with a small operation a few miles outside Donald, on the CPR main

line. In 1940 he relocated on the railway at Donald to take advantage of the booming market. Partway through the war he sold out to Brick Chapman, who reorganized the operation as Selkirk Spruce Mills. It was eventually incorporated into Evans Forest Products.

The war put the forest industry back on a sound footing, but most of it was still operating at a fairly primitive level. Trees were still felled with axes and hand saws. Although a few trucks and Cats had found their way into Interior logging shows, most skidding work was still done by horses. All the railway logging shows had disappeared. In the sawmills the situation was not much better. Most of the equipment dated back to the 1920s and had been built to cut bigger timber than was now available in most Interior regions.

Financially the industry was in good shape, but in 1945 the future looked uncertain. If the industry was going to prosper, it needed a lot of investment in new equipment. Would the armistice be followed by a slump or a boom? And if it was a boom, how long would it last? As the answers to these questions emerged, the Interior industry was gradually transformed.

Above: The sawmill crew at an unidentified east Kootenay mill in the 1930s. Slabs have been cut off two faces and the log turned by hand for the next cut. CFI

Left: Log pond behind a dam which provides water for a branch flume, BC Spruce Mills at Lumberton, 1920s. When the water was released, the logs were fed into the flume through the spillway. BCARS 73600

CHAPTER 7

Growing & Prospering
The 1940s

The end of World War II brought dramatic changes to the Interior forest industry. European reconstruction and the postwar boom in North America gave rise to a strong demand for forest products. In the first decade after the war, log production increased fourfold, while lumber production increased 625 percent. With rising prices, the value of this production rose even more spectacularly.

In response to this demand, the Interior of BC saw an enormous surge in the opening of new mills, the expansion of older ones and the consolidation of others. Perhaps typical of what was to come was the purchase of Shuswap Lumber at Canoe in 1945. This sawmill had been built by R.W. Bruhn in 1925 and sold to Ormie Harris, the founder of Shuswap Lumber, in 1937. (Bruhn also owned a pole and tie company on Shuswap Lake, which he sold to Carney Pole.) In 1945 the buyer of Shuswap Lumber was Federated Co-operatives, a Saskatchewan operation that supplied a string of co-operatively owned lumber yards on the prairies. Ten years later Federated bought another Harris mill, Chase Lumber.

In Revelstoke a local boy named Vic Camozzi opened a sawmill in 1945, after a few years' experience as a logger and portable mill operator. His new mill was in town, and when the Forest Service insisted it be given a name, he shrugged and said, "Well, it's on Downie Street, so I suppose it's the Downie Street Sawmill." The name stuck and has been retained through several changes of ownership, beginning with its sale to Federated Co-operatives in 1979.

In the north Okanagan, several small mill operations were launched that changed hands, interbred and became part of various major Interior companies over the next few decades. East of Enderby, up the Shuswap River and into Trinity Valley, Jerry Raboch built a small sawmill in 1947 out of used farm equipment and automobile parts. Raboch had come to Canada from Czechoslovakia more than thirty years earlier, as a boy of twelve, to live with his uncle. He knew no English but was an accomplished classical violinist. During the late 1920s and 1930s he worked logging on the Shuswap, and spent the war years in Europe as a member of the Canadian Forestry Corps.

> After the war, the market was real good and the bush mills were cutting ties for the CPR. There was big contracts for ties. And rough lumber, they could just get rid of it anywhere. A lot of little two-man, eight-foot mills in the bush, portables and things like that. Boundary Sawmills was buying rough lumber. There was a couple of other planer mills started up too. And then of course some of the larger or more affluent of the small mills, they put in planers eventually. That was a boom time. There were sawmills all over the place! I know we were just snowed under at Kettle Valley as far as the Forest Service were concerned. I think there was just below two hundred timber sales in that little area and they were going hot and heavy. You didn't have time to turn around. They were hard to police too. Scaling was on the honour system, although we tried to police it. But scaling for stumpage was quite an adventure for us. Many a time in the Forest Service we'd have a bad actor and we knew he wasn't turning in the proper scale, so we used to trace his rough lumber shipments. They would take truckloads over to the Okanagan and sell them to some lumber outfit over there on the side so we wouldn't have a record. We dug out the records on some of them and we'd do a log scale and bill him accordingly. It was quite an adventure sometimes.
> —Bill Uphill, who worked for twenty years as a Forest Service ranger at Beaverdell and Kettle Valley before joining Pope & Talbot, where he served as vice president until his retirement in 1990.
>
> The first Quesnel planer mill was started by H.A. Gardner at Six Mile, which is just across the river. Right in town, a guy called Manfred Hilmoe had the first planer mill on 2 Mile Flat at the end of the railway. They were all four- to six-man operations. They would buy rough lumber from the bush mills. The guy would come in with his truck, with his load of rough lumber. Usually he was tied to one of the planer mills already. By 1950 there was probably six and by '55 there was probably ten planers in town. That was the big push, right in that period, including Ketchams' in '55. They were all the same size. They came from everywhere. The Ketchams came from Seattle and John Ernst came from Nova Scotia. There were other American outfits, Pacific Western was in here. They were American but they were some person or family. There weren't the Weyerhaeusers or any of those kind of companies. They were just the little guys, and they all scrambled. Lumber was strictly a commodity like it is today, and it probably fluctuated worse than it does today. A guy would start in '48 and make a good shot of it. In '49 the market went for shit and if he didn't quite have his money together, he didn't make it 'til '50, and the next guy filled in his spot. They were all underfinanced, I think, and the market was so flexible. It would go from 45 bucks up to 70 bucks, then back down to 47 and up to 90, and if you missed one of those humps you didn't make it to the next one. A lot of the success of those guys was based on their ability to make it during that real good shot they had at the market. They had to have sufficient sales and they had to have sufficient guys tied up that were delivering to them. Before 1950 it was strictly fir. That's all you could sell. If you had spruce in your load, the planer mill kept that and didn't pay you for it.
> —Al Dupilka, a forester who retired from West Fraser Timber Company in 1995.

Page 120: Truck hauling a big load of logs in the Okanagan valley, 1952. Trucks like this one were part of the early mechanization of logging after World War II. KLM 4293

Page 121: Wally Neff feeding an electric planer at Quesnel, 1950. WFT

After the war he returned to logging; finding himself with three energetic sons, he decided on a new course of action to harness their enthusiasm. He purchased some land with timber on it, then taught his sons how to log. They sold the poles and sawlogs to Tom Malpass, a pioneer truck logger who had recently set up a mill and pole yard at Ashton Creek, near Enderby. A couple of years later, the Raboch family built their first mill. One of the sons, Gerald, taught himself a variety of mechanical skills through correspondence courses, and the family began mechanizing, acquiring trucks and Cats. By 1954 they had outgrown their site and relocated at Ashton Creek by taking over an established sawmill and planer complex. In 1962 this operation, in partnership with a Vancouver broker, William Steele, became Riverside Forest Products, which eventually took over most forest companies in the Okanagan (see Chapter 9).

Another north Okanagan sawmill was built in 1946 at Vernon, by Rex and Engel Ganzeveld. They bought a planing mill at Salmon Arm that same year. Over the next decade

Right: The mill crew at Ganzeveld Brothers sawmill, Vernon, 1952. The forklift was a recent innovation.
EDMS 1644

Below right: Simard's equipment barge on Mabel Lake, en route to Noisy Creek.
EDMS 1419

or more they bought and sold several small mills in the area—including Malpass Lumber & Poles—most of which eventually ended up, one way or another, as part of Riverside Forest Products.

For instance, Curly Noble opened a planing mill at Vernon just after the war and sold it to the Ganzevelds a few years later. After running a resort at Mabel Lake for a couple of years, Noble felt the urge to get back into the lumber business, and started a planer mill at Mabel Lake with Dave Simard to finish the lumber produced by Simard's sawmill. Simard was born near the Shuswap and was the grandson of Napoleon Simard, a pioneer logger and river driver. His father and uncle were also well-known loggers in the region. Noble had other mills in the country, which eventually ended up in Riverside. Simard himself later worked for Riverside for several years.

A number of the principals of these mills grew up together, went to school together and worked together on their first jobs logging and driving logs on the Shuswap. They worked for each other and with each other and they married each other's sisters and cousins. When the postwar boom came along, they put together various logging operations, sawmills and planing mills, buying and selling to each other and generally treating the whole business as if it were a gigantic swap meet. In the end, most of their endeavours would coalesce to form Riverside, the operation that began life up a tributary of the Shuswap River under the guidance of a concert violinist.

A few miles south of the Shuswap, at Lumby, two other companies flourished during the 1940s. Henry Sigalet, the Sugar Lake pole logger, had been broadening his interests. He acquired a number of small sawmills throughout the northern Monashee area and during the war built a larger sawmill at Lumby. In 1947 he consolidated his various operations, along with a couple of partners, under the name Lumby Timber. At about the same time, he started doing business with his son Jack, at Golden, with a pole company and a sawmill, Golden Lumber. In 1960 this mill was sold to US interests and renamed Kicking Horse Forest Products. Sigalet's Lumby mill would eventually end up with Riverside, after he sold it to Stanley Simpson at Kelowna.

The other Lumby venture of this period was started by newcomers to the Interior. Harvey and Ian McDiarmid, along with Ken Johnson, went through the war together, and when it ended they started a tie mill at Port Moody. In 1950, after operating several small mills in the Fraser Valley, they settled in Lumby and built Merritt Diamond Mills. This company spent the next fifteen years expanding and taking over several smaller mills, before selling out to Kamloops Pulp and Paper in 1965.

> I come out here in 1948 for Oliver Brownmiller. Originally, Brownmiller Brothers had the mill West Fraser has now. I had a Cat, a TD9 International, about 50 horse, I think, small compared to the machines nowadays. I was in the land clearing business in northern Saskatchewan. You couldn't do it in the wintertime, it was too frozen. The snow got deep.
>
> The Brownmillers had been logging out here for a few years and were having trouble getting the logs in. There was very few Cats in Quesnel at that time. He talked me into coming out and I worked for them in '48 and '49 and the spring of 1950. I brought a tractor with me too, and I traded it to a farmer up north who wanted to start a dairy farm. I traded it for his sawmill. Now I needed a power unit and I needed some timber. I hooked up with Brownmiller then. They had changed their operation and had a T120 Chrysler six-cylinder inline. So he said I could have the Chrysler if I wanted.
>
> This timber limit that they'd bought too with this mill had a great big gully running across the corner of it. It was not feasible to skid across the gully to the mill, so we set up there and logged it off. Sawed it in the bush and then hauled it into town. By this time Brownmillers had their planer going in Quesnel.
>
> My brother come up and went in with me. When I first started we called it B & F Sawmills (Brownmiller and Falloon) because when Oliver sold me this motor he said, Pay for it when you get a chance. Well, I said, do you want to go in on the deal? He said, That's what I'd rather do. So we went half and half and I ran the mill and he was running the other mill. When West Fraser got Brownmiller Brothers, we changed the name to Falloon Brothers Sawmill and operated 'til the fall of '69, I guess, when the mill burnt. We logged for a year, then I got out of it. I tried the resort business, and then I got back into logging again a few years later. Bob, my boy, and another lad, they had some skidders and were trying to pick up work here and there. We bought out another logging outfit that had everything—Cats and loaders, the whole bit. Called it Quadra Logging and I've been in that ever since.
>
> —Cliff Falloon, who is still active in Quadra Logging.

Consolidation of existing operations—viewed by some as a reasonable necessity and others as the voracious eating of little fish by big fish—was a process that got into full swing in the 1960s. Before that time, there was one clear case of wholesale corporate concentration. Stanley Simpson had built up his Kelowna operations before the war, through a combination of expansion and acquisition. After the war he continued to expand, with major additions to the S.M. Simpson Company mill and box plant, until he retired in 1955. His son Horace marked his takeover of the helm by buying two Peachland sawmills, Peachland Sawmill & Box, and Trautman & Garroway. Two years later he bought a sawmill at Malakwa built by Sinclair McLean, the East Line lumberman. McLean had closed one of his mills on the Fraser River and moved it to the warmer winter climate of Malakwa in 1949.

At about the same time he acquired the Malakwa mill, which included a good supply of peeler logs, Simpson built a veneer and plywood plant in Kelowna, much of the veneer going into box construction. To provide this plant with peeler logs, he took over several smaller mills with good timber supplies, buying out Sigalet's Lumby Timber in 1959. Then, true to the big-fish-little-fish scenario, the S.M. Simpson Company itself was swallowed up by Crown Zellerbach in 1965.

Quite a different sort of operation started up across the lake from Kelowna in 1951 after John and Ross Gorman were frozen out of their orchard business. Desperate for an income, they cobbled together a small box mill in Ross's machine shop, powered by their farm tractor. They started by cutting mill ends they obtained from nearby sawmills, and within a few years they were logging their own timber and operating their own sawmill, as well as manufacturing boxes. Eventually, Gorman Brothers would evolve into the Interior's major producer of high-value pine lumber.

> My brother and I had a peach orchard. In '49–'50 we had a really hard winter and we lost a lot of our trees. We survived, but the '50–'51 winter was a final thing. Another bad winter. We decided through a lot of discussions that maybe…we were both mechanically inclined and could build stuff. We were both welders. I'd taken welding courses during the Second World War and my brother was fairly mechanically minded. So we thought, let's build some simple machinery and buy some lumber from the existing mills here, Trautman & Garroway in Peachland and Federated Co-op in Salmon Arm. So we started out with $500 of capital and built our first cutoff saw and chop saw. Then we had to have a resaw, so $500 went into a resaw, or $350 I guess it was. That's how we started, in a very small way up in my machine shop, up the road from here.
>
> In those days everything was in wooden boxes—apple boxes, peach boxes, pears, prunes, apricots, everything was in a wooden crate. All from a bushel box for the apple to a small crate for peaches or prunes or apricots. That's how it all started. Millions of feet of lumber was used throughout the valley and there was hundreds of thousands of apple boxes or peach boxes or apricot boxes. Not everybody got frozen out like we did. There were parts of the valley that didn't get hurt too bad. We were at a little higher altitude and we just got nailed. We got frozen in '49, so that winter we had to start to do something. We actually started the company in January of '51 as a small box plant supplying apple boxes to the local packing houses.
> —*Ross Gorman, one of the founders of Gorman Brothers Lumber.*

West of all this energetic Okanagan activity, in the Thompson country, a couple of other major mills were established in the late 1940s. In 1946 Bert Balison started with a portable mill near Heffley, north of Kamloops. By the early 1950s he and his sons had acquired two more portable mills and set up a stationary planer at Heffley to finish their lumber. They called their new operation Balco Forest Products.

East of Kamloops, at Adams Lake, Art Holding and a changing cast of partners built a small stationary sawmill at the end of the war. Holding had worked through the war logging for Stanley Simpson. In 1945, after a fire and rebuild, he bought out his partners and pursued an expansion program, developing one of the larger Interior mills. He was also an original partner in Kamloops Pulp & Paper. Holding Lumber was sold to a coastal company, Whonnock Industries, in 1971.

The third area in the southern Interior to undergo substantial industrial growth was the Kootenays. Crow's Nest Pass Coal, which had been running sawmills to supply its mining ventures since the nineteenth century, put in a planer in 1950 and entered into the lumber business. Through the 1950s and 1960s it grew steadily, buying out several southeastern sawmill companies, consolidating its operations at Elko and changing its name to Crow's Nest Industries.

On a smaller scale, Galloway Lumber did the same thing. In 1945, Oscar Jostad and Charlie Nelson bought out MacDonald Lumber at Galloway and, after renaming it, acquired several other mills in the area. In 1970 the Nelson family bought out the Jostad interests and continued to operate it as a family business, which it still is today.

Following pages:
A state-of-the-art logging truck belonging to Pat Duke of Lumby, late 1950s. The truck is being unloaded with a steam-powered crane. NVMA

> If you were a pretty good-sized operation you'd have a little Ferguson with half-tracks on it, one-horse bunching, and that was a 6,000- to 8,000-board-feet-a-day mill. And if you just used the horses, that'll run usually around 6,000 board feet a day. We bunched out to the main trails with horses, usually a single horse, sometimes teams. Just pull 'em out and pull 'em over a log so they could back in with a tractor and slap on a chain and grab two, three logs at a time. Skidded in and put 'er on the skidway. At the mill they'd roll 'er in and it worked really, really efficient, really good. If you had a little Ford tractor you considered 6,000 to 8,000 board feet a day. If you had a bigger tractor, like a Fordson Major, then you were a big operation. You were 8,000 to 10,000 board feet a day. Everyone had 'em here. And that's what we hauled the lumber with. We had no roads, really, in this whole back country, and all we did is we'd saw the lumber, load it on a sleigh, or a wagon in the summer, hook it behind the half-track and away you went, took it down to the railroad siding. You'd never see trucks. First truck I ever seen haul lumber I thought it was a joke. It was my wife's brothers. They came up one of our arch roads with about a '53 Ford truck, brand new. It worked good but we built real tough roads. With a half-track, it didn't make no difference. It was good for us young guys 'cause we wanted to get the extra hours. We'd work in the mill all day and then at night drive a half-track. We'd haul lumber all night, down to the siding. We really enjoyed it at night when it's thirty-below out there, the northern lights, and you're coming down, the ice spraying off your tracks and a load of lumber in line, it's a real nice life, eh. Really nice.
> —*Bruce Kerr*

The venerable Rogers sawmill at Creston was taken over by Don Burns and Alf Farstad in 1945. Farstad started out as a logger at Duck Creek, north of Wynndel, in the 1930s. In the late 1940s the men started Cranbrook Sawmills, and spent the next decade building up both companies, including a plywood plant at Creston, before selling to Crestbrook when it was formed in 1956. Farstad, who became Crestbrook's first president, was well known for his love of gambling and is reputed to have acquired more than one timber claim at the poker table.

Starting from scratch with a tie mill at Radium, Lloyd and Earl Wilder gave up their Alberta trucking business to join the BC forest industry in 1945. After building and running a stationary mill at Fairmont for a few years, they moved it to Radium. They went into buy-out mode and picked up several small mills in the Canal Flats-Radium area. They sold the whole works to Revelstoke Sawmills in 1964, and from there it was eventually passed along to Slocan Forest Products.

> I came to Canada in 1951, to Telkwa. And at that time I started as a canter man at the sawmill. At that time they call it TF&M. When I started we had quite a few mills here. You had about twenty-five, maybe thirty little sawmills which would cut in the neighbourhood of 4,000 to 8,000 board feet. They sold the lumber to TF&M here and to them two planer mills in Smithers, most of it. When I was sawing for the company they cut about 20,000 board feet a shift. We were relying on very poor timber that came out of the Telkwa River valley, old balsam, and the recovery was very poor. At that time chips were not made, so everything went into the fire. We were cutting not just two-inch random, but four-by-fours and four-by-sixes and six-by-eights, and all kinds of stuff like that. If we had the quality of wood, then of course we were cutting lots of two-by-nines, which was a pretty good price at that time. And lots of ties—the two-inch lumber was more or less a by-product of all the other stuff we were cutting. At that time there was a lot of waste because a lot of poor lumber came in. And it was not on account of the timber, it was on account of really poor manufacturing done with the little sawmills. Some loads, they figured they had 6,000 or 7,000 feet of lumber on it. After it was trimmed, turned out to be only 3,000 feet. And a lot of people complained. At that time a lot of lumber which had been put through the planer, they had to put it through twice in order to get the right size.
> —*Fritz Pfeiffer, who has logged and operated sawmills at Telkwa since 1951.*

Alf Farstad (right) hauling slabs in the east Kootenays, 1932. Farstad later became president of Crestbrook Forest Industries Ltd.

Bottom: Members of the Kalesnikoff family in the mid-1940s. Left to right: Pete W. Kalesnikoff, Larry Postnikoff, Sam S. Kalesnikoff, Sam W. Kalesnikoff, Koozma Kalesnikoff, Mike Kalesnikoff (in front), John Kalesnikoff, Peter Kalesnikoff. From its beginnings in the Depression, Kalesnikoff Lumber Co. Ltd. has grown to become one of the most respected and well managed logging and sawmilling companies in the southern interior. Peter Kalesnikoff (on far right) is currently the president of the company.

These mills were only a few of the operations which got off the ground in the southern Interior during the first decade after World War II. In total there were probably several hundred sawmills operating throughout the southern third of the province during this time. Most of them were small, portable operations that in time faded away or were bought up by bigger players in the industry. Milling lumber was a fairly precarious business at this stage. While demand and prices were good, they were erratic. And obtaining reliable, experienced workers was exceedingly difficult.

The postwar labour shortage dramatically affected all the forest industries, perhaps more than other economic sectors. The entire economy was booming and there were jobs everywhere, so loggers and mill operators faced stiff competition for workers. This situation had two immediate consequences for the industry.

The first was the migration of the woods union, the International Woodworkers of America (IWA), into the BC Interior. The IWA evolved out of the radical Industrial Workers of the World (Wobblies), a significant political force throughout Washington and Oregon after its formation in 1904. The union was formed at Tacoma in 1937 and for the next few years was riven by left-wing factionalism, culminating in the purging of its communist members. By the beginning of World War II, the BC wing of the IWA consisted of only a few hundred members on the Coast.

During the war, prices and wages were fixed by the government, and after lumber was declared a strategic war material, movement of workers in and out of the industry was restricted. Although the union renounced the use of strikes for the duration of the war, there was a two-week strike on the Coast in 1943 which resulted in the BC Loggers' Association having to accept the union as the industry's bargaining agent. By the end of the war the union had 15,000 members.

> George McGuinness, that I worked for, he still used teams on his poles. He took out a lot of power poles. At that time they also used a lot of fir and larch if it was straight enough, and the CPR was the only one at the time that bought pine poles, to my knowledge. But he had contracts with them all and sold a lot of stuff to Galaway and Carney Pole. They kind of put the damper on fir and larch for a while there because climbers' spurs wouldn't dig in and there was climbers getting hurt. He used a lot of teams skidding right out of the woods if the ground would allow 'em, and he'd subcontract. He hired hand fallers even after the power saws were in, because of the delicate nature of the poles. They'd have a set of hand fallers—crosscuts—and a teamster. The men that was doing the falling would cut the stumps very low, and grub hoe out to leave it so that you could get through with a team. The price at that time was a nickel a running foot for poles; that's what he paid the set of fallers. They usually were partners that would take the contract. The team was supplied by George and they got a nickel a running foot, but they had to pay the wages of the teamster out of that.
> A single-axle truck with a single-axle trailer hauled roughly a thousand lineal feet. And they could get a truckload in a day quite nicely. A teamster's wages at that time was $1.30 an hour, but he got nine hours a day because he had to look after the horses.
> —Don Mallard, who started logging and sawmilling on his father's ranch at Fort Steele in 1950, and has worked in the business ever since.

After the war, the wage freeze continued and even though mill owners, especially in the Interior, tried to get wages raised to attract workers, they were turned down by the federal government. This ruling angered workers and owners alike. At the East Line's Sinclair and Eagle Lake mills, Roy Spurr and Don McPhee advised their employees to get organized and bring in a union. For its part, the union had for some time been engaged in a bitter factional dispute over who would be sent to organize in the Interior. In 1945, Ernie Dalskog, a tough, experienced organizer and a leader of the union's left-wing faction, travelled through the Interior, setting up locals at most of the bigger sawmills.

The Woodworkers Industrial Union of Canada was a communist oriented breakaway faction of the International Woodworkers of America. Left to right: Hjalmar Bergren, Harold Pritchett, Ernie Dalskog, Jack Forbes. Dalskog was the IWA's leading organizer in the Interior.
UBC/Pritchett BC 1529/23

Unlike their counterparts on the Coast, Interior loggers did not rush to join the union. They did not work in big camps, as did many coastal loggers, and most of them logged only in the winter, working on their farms or at other seasonal jobs in the summer. As a result, Interior union locals were built around the sawmills, and millworkers dominated the union.

In 1946 the IWA called a province-wide strike. The government was maintaining wage controls and ruled there could be no more than a 5-cent-per-hour wage increase. The union asked for 25 cents; the employers offered 5. After five weeks, the government stepped in and took control of the Interior box plants so that the fruit crop would not be left to rot in the fields. In the end, the workers gained 10 cents an hour and a forty-four-hour work week, and the IWA gained 10,000 new members. The next strike, in 1953, almost wiped out the IWA. This time the Coast and Interior locals negotiated separately. Coastal workers endured a forty-two-day strike, but the Interior strike was staged during a dispute within the union and lasted more than three months, until just before Christmas. In the end, the union got only a small raise in wages, but lost hundreds of members, its cash reserves and the support of the general public.

I drove horses a lot of the time in the winter. If you had a sick horse, you nursed him that night and maybe all night. You worked right 'til Saturday night and you fixed harness on Sundays, ready to go Monday morning. Overtime was unheard of. I never heard of it. At the mill in Parson we used to saw in the daytime and plane sometimes from six in the morning 'til eight o'clock, get breakfast, back in the mill for the day and then after supper you might go back to the planer. You worked an hour and a half, you got an hour's pay. You worked an hour and three-quarters, you got an hour's pay. If you worked two hours, you got two hours' pay. That was 80 cents. Time and a half meant you worked an hour and a half for an hour's pay.

I think about 1948 or '49 the union started coming in. The old fellow down there couldn't stand it. He made them get out of the bunkhouses because if they weren't gonna work they weren't gonna live there. It was a new thing that he just couldn't understand. I myself, when I heard that they didn't have to work Saturdays, wondered how the mills were gonna make it now, and how the fellows were going to make enough money to live.
—*Orond Brashier*

When I came to Prince George in 1951 the East Line was still the powerhouse of the industry. But it was developing rapidly and as it developed, all these portable mills in the bush were bringing truckloads of lumber in here and it was all being remanufactured on planer row. That's where these high piles came from. They would buy the rough from these portables out in the bush and they would dry-pile it in these high piles all winter. In the spring they would take it all down and plane it and chip it. So you had primarily all your raw material to remanufacture. The rough lumber was being hauled in in the wintertime. It was stockpiled. They did manufacture some, but millions of feet were stockpiled down there. Basically, what they were doing was air-drying it. By air-drying it they got the weight down and by getting the weight down they could ship more lumber on the cars and their shipping bill was quite a bit less. With that development and with all these bush mills springing up all over the place, there was more and more lumber coming into Prince George and into these planer mills. They were shipping more lumber than some of the big boys on the East Line. So the balance shifted.
—*R.J. Gallagher, manager of the Northern Interior Lumber Association from 1952 to 1972.*

The mill owners had their own associations. The first to be formed, near the turn of the century, was the Mountain Lumbermen Association, composed of companies from the Kootenays, Kamloops and southwestern Alberta. It folded during the Depression. During World War II, a second southern organization, the Interior Lumber Manufacturers Association (ILMA), was formed by eight mills in the Thompson–Okanagan area. It later grew to represent companies throughout the southern Interior. Later, when the Cariboo

Piling rough lumber for drying at the Grindrod Lumber Company. When dried, it was planed to its final dimensions. EDMS 3948

Lumber Manufacturers Association was formed in 1958, some ILMA members moved over to that organization.

In the north, a division of the Canadian Manufacturers Association, the Northern Interior Lumber Association (NILA), was formed at Prince George in 1940. Thirty years later this organization severed its relationship with the CMA and affiliated with the BC Council of Forest Industries. In 1952, NILA hired a new manager, R.J. Gallagher. When he first visited Prince George, Gallagher was employed by the BC Health & Welfare Department to visit employers and explain the payroll deduction system of the new hospital insurance program.

He stopped in at Rustad Brothers Planing Mill and, all dressed up in suit, tie and well-shined shoes, carefully made his way along some planks laid across the muddy mill yard to the office, where he was met by Mel Rustad. Gallagher explained the deduction system to Rustad, who listened closely without comment. Then, as Gallagher got up to leave and opened the office door, Rustad grabbed the natty civil servant by the scruff of his neck and the seat of his pants, and hurled him head first into a big mud puddle at the bottom of the steps.

Unfazed, Gallagher picked himself up, marched back into the office and informed Rustad: "You owe me a drink for that." Equally unfazed, Rustad reached into his desk drawer and pulled out a bottle of rye. By the time it was empty, Gallagher had been offered a job as NILA's first manager. He accepted and ran the organization for twenty years. These were the sort of people who sat across the negotiating table from the IWA in Prince George.

It took us years to get a grade rule. We had more meetings back east because we had to deal with the Maritimes, they had their own grade rules, the Canadian Lumberman's Association had their own grade rules and they covered Ontario and part of Quebec. Quebec had some grade rules of their own. Nobody was going to change the system that they had for their people. And the Canadian Lumber Association was all-powerful. Their offices were right in Ottawa. So we used to have to go down there and then the boys from Quebec would come in, the boys from the Maritimes would come in. There'd be ourselves, the southern Interior, Alberta, Saskatchewan, Manitoba and the Coast. And we'd all go down there and try to hammer out grades. A lot of our big spruce up here, they didn't know what spruce was back in that time 'cause it was so small, and we used to have a lot of trouble on the top grades and the wide widths. That was a real tough thing. In the Maritimes, and in eastern Quebec too, they had a type of pine, beautiful wood, and that had to be accommodated because it wasn't like our lodgepole pine up here or it wasn't like our jack pine. Their pine was a different pine, so it had to be accommodated because it was a decorative wood. You could use it for inside panelling and stuff like that. Now, our spruce was fine. If it was properly dried, it made excellent panelling. But it had to be properly manufactured, properly dried. They became more cognizant of the way they were making their product and the attention they were giving it when it came to drying. They used to ship lumber PAD they called it—partially air-dried. Some of that stuff came right out of the bush and into the rail cars, PAD. It must have taken us over five years to get the industry as a whole to accept the rules. We had to accommodate everybody and we had to have one rule book, you see.

—*R.J. Gallagher*

The rubber-tired arch, developed in the 1920s, enhanced the skidding capability of the crawler tractor. Arches came into wide use in the Interior in the 1950s, as in this West Fraser logging show in the Cariboo. WFT

The second consequence of the postwar labour shortage was the mechanization of logging. Throughout North America and western Europe, logging companies were in the same bind. The war had accelerated industrial development, and when it ended there were a lot of high-paying jobs available that did not involve the hard work and dangerous conditions that were part of logging. In the southern USA, black loggers left the hot, humid forests in droves to work in air-conditioned factories. In the north, where logging was still done in the cold winter months, with hand saws and horses, for relatively low wages, workers had the same response.

Everyone in the industry knew the solution was simple and straightforward. Using conventional tools, a logger could not produce enough in a day to justify paying him the wages he needed to stay at work in the woods. But if he had better, more productive tools, he could be paid more and persuaded to stay at work in the woods. Some people believed that mechanization was a means of reducing the workforce. Although this reason was probably a factor in subsequent mechanization phases, when labour costs had increased to unsustainable levels, in the postwar period the last thing a forest company wanted was to get rid of good workers. The logging industry did not operate in isolation from the rest of the economy, and if loggers could not be provided with the right tools, they would end up working elsewhere.

One of the first machines to be adopted on a large scale after the war was the crawler tractor—the Cat. It had been used occasionally in the Interior before the war, but in the late 1940s it was a much improved machine. It cost less and it was widely available, as were parts and trained mechanics to keep it running. Also, loggers who could work with horses were in short supply. Farms had begun to use tractors, so this training ground for teamsters had disappeared, as had an important source of the horses themselves.

My first job up here was loading and hauling logs with a truck and trailer from here to Robson, in 1940. It was a big thing in those days to have a three-ton truck, and a trailer made out of the rear axle of some other old truck. The heavy-duty three-ton Chev truck in those days was called a Maple Leaf. Single-axle hydraulic brakes with a vacuum assist. The trailer was made out of another old truck axle that had mechanical brakes on it. You had an extra brake handle in the cab with a clothesline wire running back along the reach to the trailer brakes. Going around a corner to the left, you'd have to pull the handle right back to the seat, and going around a corner to the right—why, you probably wouldn't be able to pull it at all, and the brakes would start coming on themselves.

Those poor little trucks with less than a hundred horsepower would drag loads of logs up these hills. And, snowing, you'd be putting the chains on. If you left them on when you got over the top, well, you'd beat them all to pieces going down, so you took them off. Come to the next hill, you'd put them back on. It was quite a game. But we really thought we were doing something in those days.

It's scary to think of how we loaded the trucks. You'd have a deck of logs above a bank held by chips and rocks and one thing and another, and you'd put skids from the bank onto your truck and two men would get up there with peaveys. Of course, you wore caulk shoes and stood on this confounded skid and you'd start the log coming down. You'd hold it at the bottom of your reach, down as far as your arms would go, and the other guy would jam his peavey in and hold it up there so that you could unhook yours and he'd let it roll down. Well, there's nothing in the world to stop this thing from slipping. There's snow and sleet coming down and those skids are slippery, and you let the first few logs on that way—if you're still alive.

—*Bob Cunningham, who was born in Crescent Valley in 1918 and drove logging trucks in the Kootenays for many years.*

A Baird Brothers Logging truck crossing Cooke Creek east of Enderby, 1952. EDMS 90

At the point the first Cats came in, there were a few Cletracs and a few Oliver OC3s, and we thought they were real monsters. Of course you'd put your foot in front of 'em and stall 'em, they're just toys. But they did the job. I built roads with OC3s. We figured they were a pretty big Cat but they couldn't skid the production that half-tracks and the horse combination could. The first Cats to come into this area were mostly John Deere. There was a local fellow owned the dealership here, so everybody went John Deere. We started out with the little MCs, the 40s, then the 420s, 440s, and worked way up to the 450. Then all of a sudden the other lines came in. Allis-Chalmers started making a mark with their HG5s, and Caterpillar with their D4s, and International TD6s and stuff. They were fairly big Cats to us. Then it all of a sudden just ricocheted, eh. Everything got bigger. We went into arch trucks, skidding with arch trucks. Freezing roads in, then use the big arch and behind it you take forty trees or so. You give a push with a Cat to get 'er started and away they go. Head to the landing. We brought in the first Timberjack skidder. We had the first one in the area here. It was quite a funny-looking thing but it worked good. Then we bought a bigger one and we started arching with it and it worked good for us. We always started new things.

—*Bruce Kerr*

Cats were good for a skidding distance of a mile or so. This seemed like a great distance to an old horse logger, but timber within a mile of a river or a road was becoming scarce. What was needed was a machine to haul logs beyond the range in which a Cat was economical. The railways were finished as logging machines, the terrain and timber stands which could support them having been logged already.

Fortunately an ideal machine was readily available, in the form of surplus military trucks. A wide range of vehicles came on the market right after the war—two-wheel-, four-wheel- and six-wheel-drive trucks equipped with powerful motors and built to handle tough conditions. They were snapped up by Interior loggers, who installed winches on the rear decks and arches. With one of these rigs, workers could yard logs over short distances to roadside, top first, the leading ends hoisted off the ground and skidded to the landing or the mill. On a well-prepared road of packed snow, one of these arch trucks, as they were called, could skid a "turn"

A six-wheel-drive military surplus truck. These trucks, with the addition of a winch and arch on the deck, were immensely popular in the Interior from the late 1940s until the advent of rubber-tired skidders in the early 1960s. This one worked at McAmmond's sawmill at Clinton in 1962. EDMS 2717

of a dozen or more logs several miles at phenomenal speed, and return for another load even more quickly.

With the addition of an A-frame, such a truck could be turned into a "jammer" that could yard logs several hundred feet out of the woods. After this introduction to cable logging, it was a natural step for some loggers to move into high-lead systems, using portable spars and specially built logging winches, such as the Lawrence 10-10 donkey engine, built in Vancouver for coastal loggers. The problem with this system was that Interior timber was generally small and cable logging systems are more efficient on bigger logs. However, the use of arch trucks, jammers and high-lead machines did introduce the concept of tree-length logging to the Interior at this time.

Perhaps the greatest mechanical boon to Interior loggers after the war was the chain saw. A great deal of the development of this machine took place on the Coast, just before and during the war. By the late 1930s, the steady stream of skilled Scandinavian hand fallers had slowed to a trickle and coastal logging companies in BC and the USA were finding it difficult to obtain fallers. The BC Loggers' Association launched a campaign to find and develop suitable power saws to alleviate the effects of this shortage. In 1936 it obtained some German-made Stihl saws and set up a program to recruit and train power-saw fallers. When the war started and Stihl saws were no longer available, a Vancouver company, Industrial Engineering Ltd. (IEL) began building replicas.

Early chain saw falling at a Holding Lumber show on Adams Lake. BCARS 73909

My brother Ozzie brought in a lot of these big ten-ton Macks, six-by-sixes, and he put these arches on and sold them. We would sell them around this country and we had to take some of them back, so we started arching. That's what we used at Mabel Lake from '58 to '62, or something. We used two of them on that job there. When the big gas motors went out of them, we put in great big diesels and we had to push the radiator out about a foot and a half to get these motors in. Boy, they could skid then! If one broke down sometimes up at Noisy Creek, I'd run the other one at night shift. It ran right around the clock. You could see those big exhaust pipes glowing scarlet.
—*Audrey Baird*

Everybody here used power saws [by the early 1950s]. It was already the time for one-man power saws. McCulloch sold quite a few here. Pioneer had sold quite a few. Then you had Titan and Homelite, Clinton. I had some of them. Some of them made a lot of noise. If you wanted to fall a tree you had to turn the blade. And on some of them, the Clinton for instance, fuel injection was by pressure. The tank had to be under pressure in order to get the gas if you turned the saw. And you could saw just like with an ordinary saw today. So you can imagine how much trouble you had at that time if the pressure wasn't right. Some of them had belt drives, some of them had gear drives, so they caused a lot of problems, especially if you used them for falling. The limbing was all done by axe, a double-bitted axe.
—*Fritz Pfeiffer*

The first power saws weighed more than 100 pounds and required two operators each. Old-time hand fallers, who knew how to fall trees, were not willing to pack these monsters through the woods. Nor did they know how to operate and maintain the gas engines which powered the saws. The solution was to hire husky young men who had grown up on prairie farms and learned how to run and fix machines, and pair them up with experienced fallers. The machine men, as the young recruits were called, carried the heavy end of the saw, while the old hand fallers took the lighter end, a handle on the end of the cutting bar, and managed the actual falling procedure. Eventually, as the saws became lighter, IEL and other companies produced one-man saws and the prairie recruits, having by now mastered the art of toppling trees, took over a large portion of the falling business. Two-man saws were not widely used in the Interior as they were too cumbersome to operate in the denser stands of smaller timber. More manoeuvrable one-man saws, such as the IEL Pioneer, the Titan Junior and the McCulloch, took the Interior logging industry by storm during the early 1950s.

The second chain saw to appear in Enderby, operated by Audrey Baird (left) and Mike Ludwig, 1946. The Bairds later acquired the McCulloch chain saw dealership in Enderby. EDMS 4311

I got a job working for Waldie & Sons up in the Arrow Lakes, falling. It was some of the first fancy power saws, ones that were up to date. The old PMs, they had reed valves in them. And they weren't cheap at that time. A new PM power saw was $407. You bought the power saw, you invested $407 in this piece of machinery. Plus your gas can and your wedges and your hammer and your axe. And your lunch bucket as well. And your caulk boots and your hard hat and your fire extinguisher. Now you had to make enough on contract to pay for the saw and still try to have a family and grub, and your transportation by Greyhound from Nelson to Nakusp. Then hitch a ride from there down the lake to Arrow Park, cross the ferry, fourteen miles up to Caribou Lake.

The best I ever done when I was at Waldie & Sons was somewhere around $600 a month. And that's not working eight hours, like they do nowadays. That's twelve hours. There was no limit, because I was under contract. I could go right after breakfast and work right through 'til suppertime, and after supper I'd go out 'til it started to get dusk.

In 1952 was the last year I worked there, because Canadian Cellulose was taking over officially in '53. They were taking Waldie's over, and of course the prices were dropped. The contractors that were working for Waldie & Sons were all called in, one at a time, to Castlegar, to the office. There were no more advances. You got paid at the end of the month for what you produced, and the price was dropped $2 a thousand. So I decided I better look elsewhere and I went to work for Passmore Lumber Company.

Waldie, he was a little different. He was an honest man and there was very seldom a contract signed between a contractor and Waldie & Sons. It was a handshake, the old way, and that was it. Waldie would go up the lake once a week or so in the tug and meet all the contractors. He would give them so much, whatever he had, because money was short. When Celgar came in, this was different. If you could not finance yourself for thirty days, there was no advances in between.

—*Joe Wrangler, woods foreman for Kootenay Forest Products until it closed. He was born in 1919 and worked falling and driving logging trucks for many years.*

Brownmiller had a couple of guys one time that come in from the Coast, a two-man falling crew, and they had one of these big IELs [Industrial Engineering Ltd.] with a four- or five-foot bar. They only lasted about three or four days. They happened to be in kind of small timber at that time and they had a great big wide bar on them. Well, these two guys would get set up and if the tree leaned back a little bit, the saw would cut through the tree before the back of the bar got into the cut. The tree would lean back and pinch the chain and the saw would back up and set this guy on his ass. They were fighting all the time, one telling the other guy, "Why didn't you hang on to the bloody thing?" It'd knock 'em down, and the snow was deep that winter. Holy lightning! They lasted about three days. They had been bragging about how they could fall when they come but I would say, sure, if we'd twenty- or thirty-inch trees so they get the bar in partways and then get a wedge to hold the tree up, all right. But see, you couldn't wedge that saw in these small trees. And then they'd have to chop it out. Well, that's a dangerous thing too. That's a widow-maker for sure. Now and again they'd hit the chain with their axe.

The first saw I bought was a disaster. It had an English motorcycle motor in it. When you tipped the saw over, you had to tip the carburetor up to keep it straight. It was a lost cause. The guy that invented that saw didn't have too much between the ears. And I had less for buying it!
—*Cliff Falloon*

Among the early recruits to the 1936 power-saw training program on Vancouver Island were two brothers from Alberta, Tom and Dave Ainsworth. They worked at Caycuse for the duration and set up their own contract falling company when it ended, working at various logging camps on the southern coast. A lot of fallers were injured or killed in those days, and the Ainsworths soon decided to get into a different line of work while they were still in one piece.

They bought a portable sawmill and tried to make a living with it by custom-cutting lumber in the Fraser Valley. But most of the timber there was too big for the saw, so in 1952 they headed for the Interior. They ended up with a job cutting lumber for Clinton Sawmills. When that ended they moved to another job at 100 Mile House, settling in there and adding a portable gang saw—probably the first one used in the Interior—and a planer. By the late 1950s they had moved to a permanent site on the railway, put in a dry kiln, obtained their own timber sales and set up their own logging operation.

David Ainsworth (right), on the falling end of a Reed Prentice saw, and a co-worker, 1942. Ainsworth, a pioneer faller, later founded Ainsworth Lumber. Chain saws were developed on Vancouver Island during World War II. Ainsworth collection

When the strike was on in '52, we had the mill operating at Jones Creek or Bridal Falls, someplace there. We were tied up until the strike got settled so we drove up to the Interior, just north of Williams Lake, to Jorgenson and Wells. They were logging for Western Plywood, long before it became Weldwood. They had a veneer plant for peelers at Quesnel and they took the sawlogs and farmed them off any way they could. They'd accumulated huge piles of fir logs and they were pretty good logs. They were skidding them in with D7s. They didn't even use arches, just dragged them on the ground. And when they got done whittling the peelers off, they took the blade and pushed them up in a pile and drove right up on the piles. They had piles as big as this building, built with a Cat, gravel and all mixed in with them. We really wanted to go and try it, but we couldn't figure out how we'd ever be able to handle it because the logs had grit and gravel ground right into them. We came back kind of discouraged, and when we came south of 100 Mile, near Clinton, there was a planer mill called Clinton Sawmills. The superintendent from the mill and a couple of fellows were standing around talking and we told them what we were doing, and they suddenly became quite interested. "You might be just what we're looking for. We have a fellow with a brand new D4 Cat and he's logging at a mill out towards Meadow Lake. He has a brand new machine and he's pretty ambitious. He's logging for a mill that isn't cutting enough to keep him busy and he says maybe we could put a second mill in close by. He could log for one in the morning and one in the afternoon. It would keep him happy and we'd like to keep him."

So they suggested we go out there. They said, Drive out that road and you'll see a sign, just a board on a tree, and it says "A Friend," that's where this guy is. It turned out his name was Albert Friend, the guy with the mill, it was kind of poetic. We got in there to this mill and, lo and behold, the guy was running this mill with a WD9 tractor, just like ours. So we're standing there on the landing and this new D4 comes chugging in there and dropped the turn of logs. The guy on the Cat was Stan Halcro.

We had a look and went back and asked these guys, "How much are you paying?" They said they'd give us $20 a thousand for the rough lumber. So we started there and discovered to our amazement old Albert Friend with his mill that had the same power as ours was cutting more lumber than we were. It took a little while to figure out the Interior logs and really get the hum of things.

—David Ainsworth, one of the founders of Ainsworth Lumber.

By 1960 there was not a lot of fir or spruce left in the area, so the Ainsworths took a big gamble and started cutting lodgepole pine. There was still no demand for this wood. It was small and considered a weed species, useless for industrial lumber. But it was plentiful in the BC Interior, especially in Cariboo country. No one had ever tried to market the stuff—except as railway ties earlier in the century—until the Ainsworths came along.

The Grindrod Lumber Company and its new lumber truck, 1946. EDMS 1642

They began producing kiln-dried, packaged, precision-end-trimmed lodgepole pine studs and selling them through a Vancouver lumber broker. They printed a fancy brochure and painted the ends of their studs a distinctive green colour. They worked hard to fill their orders on time and marketed their product as something special. It worked, and before long a couple of other Cariboo mills—Pinette and Therrien at Williams Lake and the Two-by-Four Stud Mill in Quesnel—began cutting lodgepole pine. And then everybody in the country was into it and an entirely new sector of the industry was established.

A number of events occurred in the early postwar period that would heavily influence the future of the Interior industry. The first was set in motion long before the war ended. In 1941,

> My father, Gabe Pinette, and my uncle, Dillard Therrien, and Roger Tharion, my cousin, moved a portable mill from North Vancouver to Williams Lake in the late 1940s. They heard about the opportunities in the Chilcotin and the Cariboo so away they went and moved the portable mill up to the Interior. They ran out of gas and money in Williams Lake, and that's how they got there. If they'd had a little bit more gas, a little bit more money, they probably would have ended up in Quesnel or Prince George and they'd of had better timber to work with. But they were able to establish themselves and pick up licences or work with people with licences. In stages, over the next few years, they set up a planer mill which today is the Timber West site. They were doing custom processing at that site for a number of other portable mill operations, and the business just grew from there. Ultimately the portable mill was relocated at the same site so it became a permanent sawmill. Then the opportunity presented itself to get into the stud business with lodgepole pine. Initially a portable stud operation was established in the Chilcotin and then that was also centralized in Williams Lake and grew with the business there.
> —Conrad Pinette, president of Lignum Ltd.
>
> In 1955, just before the end of the year, we bought a planer and set it up with our portable mill. But by this time we'd added a gang saw, sash type. I think we put the first gang saw in a portable mill. It was on skids with a power unit on it too so that we could load it on the back of a lumber truck and move it. And it was driven with a Ford V8. And by this time we had traded off the tractor and had a GM gas engine which we built a gearbox for. Much better than the belt drive. A whole bunch of people adopted that system of driving portable mills when they saw how easy it was.
> —David Ainsworth

> Lodgepole pine attracted my father and it also attracted David Ainsworth, and the two, independent of one another, established their respective stud operations in Williams Lake and 100 Mile. It was considered a weed species and it was not part of the inventory at the time, so they felt that it had some traits that would lend itself to being turned into studs and be marketable. Douglas fir studs were quite dominant. A lot of plywood operations took the cores and made studs and were selling them in the US at that time. So they had to put in an effort to properly dry the pine studs, and that's when wood kilns were established. Then a lot of one-on-one development of customer relationships.
>
> I was actually going to do my university thesis on establishing a weed species into a marketable product but I got so busy running the business I didn't have time to do it. From a practical standpoint, that's effectively what we did, in conjunction with Ainsworth. Both of us were working very, very closely in those days with a company called Battle and Houghland. Tucker Battle and Harold Houghland were two marketing-oriented people and they had a marketing and sales force based in Vancouver. They helped finance both the Ainsworth operation and our operation in the formative days when we established in Williams Lake and 100 Mile respectively. It's their people, in conjunction with ourselves, that really made the market thrust. They convinced a lot of established consumers in the US marketplace to take on these studs to diversify their product mix.
>
> The big coastal companies wouldn't even have believed that lodgepole pine was a tree, let alone an acceptable species. They were logging Coast hemlock, Coast fir and Coast cedar—very, very large trees. They were probably destroying more underbrush that was of a larger size than any of the lodgepole that we and Ainsworth logged. It was about eight inches in diameter when we started, which is larger than what we're milling today on average. We're probably milling six-inch diameter lodgepole pine today, but in those days the trees that we were harvesting were probably eight to nine inches.
> —*Conrad Pinette*

Chief Forester E.C. Manning was killed in a plane crash and replaced by C.D. Orchard, a former Prince George district forester. Within a year, Orchard had sent a confidential memorandum to the Coalition government, proposing dramatic changes to the provincial Forest Act. The government decided to adopt these changes, but first appointed a Royal Commission to "educate" the public and the industry about its intent, and to receive their opinions. Chief Justice Gordon Sloan undertook the job, holding hearings throughout the province and publishing his report in 1945.

Sloan's report recommended numerous changes in provincial forest policy, which centred on two broad areas. The first was to regulate the rate of timber harvesting to assure a steady supply of wood, and thereby a permanent forest industry as a solid foundation for the provincial economy. The second was to establish a form of forest tenure to enable and encourage the growing of timber crops for future use. In response to the question of tenure, the government took Orchard's advice and passed legislation in 1948 creating the Forest Management Licence (FML), a long-term lease on government-owned forest land granting timber harvesting rights and responsibility for future forest management.

A small mill operating at Halfway on the Mabel Lake Road east of Enderby. The mill was typical of Interior milling operations between the 1920s and late 1950s. The cut lumber in this photograph was probably trucked to Enderby for planing.
EDMS 1252

Orchard believed—somewhat naively, it turned out—that FMLs would induce hundreds of established forest companies to expand their horizons and get into the business of managing forests to provide their future timber supplies. What he did not foresee was that the granting

> Soderburg's camp was up the Salmon River valley about fifteen miles from Salmon Arm. The timber here was the dry belt ponderosa pine. These trees grew to a very large size—trees of up to four feet in diameter were quite common. The area they grew in was quite steep but open with bunchgrass growing. A sort of range country. In fact there were a number of cows around grazing this grass.
>
> Power saws were becoming quite popular. Ted Soderburg had a Disston, the only saw I have ever seen that had a straddle chain. This means that instead of a bar with a groove for the chain to run in, the bar was without a groove and the chain was made so that it ran straddling the bar. The engine was of high compression and had a lot of power although the saw itself was heavy.
>
> Skidding was done by horses. Being that the sidehills were open and steep, one had to be an excellent teamster. Sometimes the log that was being pulled would start to roll downhill, and at this very moment the team would have to be pointed uphill to prevent a tanglement of wood and horses. Other times a big limb would penetrate the soil as the tree fell, and in no way would the horses be able to pull the log free. A chain would be attached to the log in a manner that would roll the log, exposing the limb that had to be cut off.
>
> The only trees we were allowed to log were the ones that the Forest Service had marked. These were marked by a blaze with an axe on the standing tree and also on a root of the same. The area blazed would then be hit with a stamping hammer, leaving a mark that consisted of numbers. They usually left many nice clean trees to go to seed. Some of the numbers were almost impossible to read, so we took the chance sometimes and blazed a nice big tree ourselves and hit the blazed part with the back of the axe. Maybe we fooled the Forest Service, I don't know. Anyway we did not get caught. This was the first selective logging that I was involved in.
>
> —*Walter Anchicoski, who was born in Enderby in 1929 and spent most of his working life logging throughout the BC Interior. He is retired in Prince George.*

of long-term, secure cutting rights at no cost amounted to bestowing an immense capital gain on anyone who received a FML. The government was flooded with applications from existing BC forest companies, as well as from large foreign investors.

Even before the legislation was passed, the government granted the first FML to a large New York firm, the Celanese Corporation, which had never conducted business in BC and which proposed to build a pulp mill at Prince Rupert, in Forest Minister Edward Kenney's riding. The Columbia Cellulose FML covered a huge area of more than 6 million acres (2.4 million ha), stretching far into the Interior up the Skeena and Nass river valleys. The far end of the licence was 250 miles (400 km) from the mill.

From the outset this was a misconceived proposal, and typified the problems inherent in the government's new tenure system. The mill had been designed to produce acetate for yarn production, not pulp for paper production. When the anticipated acetate market failed to materialize, the mill had to be converted at great cost. The company had intended to drive

> Those days a person could quit a job and be working somewhere else the very same day. While in Enderby filling up my car at Baird's garage, I heard that they—Baird Brothers—were to take out some railroad ties up at Hidden Lake and were in the need of a falling and skidding contractor. I applied and got the job.
>
> After getting a taste of the power saw, the crosscut was out. Baird's garage sold the McCulloch while across the street Prichard Motors were agents for the IEL, later called the Pioneer, which were cheaper, so I bought an IEL. That saw had a hand clutch which had to be engaged for the chain to move. The clutch itself was a number of caulks which when worn out could be replaced at a drug store. The carburetor had to be upright to work so a little pin was pulled and the gas tank and carburetor were moved to the upright position. A very crude saw, but a lot better than the old crosscut saw. And if it broke down, which was often, a person had to go to town to fix it, taking at the same time the opportunity to down a few at the local establishment and catch up on the latest from other loggers that were usually there.
>
> —*Walter Anchicoski*

GROWING & PROSPERING • THE 1940s

A heavily loaded truck hauling into Upper Fraser Mills, 1949.
PGL

logs down the Nass and Skeena rivers, but when that failed, an expensive road system had to be built. And for years, instead of making the best use of the high-grade spruce found on its FML and building the sawmills needed to cut the valuable lumber available, Columbia Cellulose ran these logs through its chippers and converted them into pulp. Interior communities in and around its FML were thereby deprived of the economic benefits to be had when sawlogs are milled into lumber and the waste is made into pulp. Fifty years later, this operation was still an economic albatross to the people of the northwest Interior and the economy of the province.

Bill Jones and Bill Demaid owned the tie mill that we were to supply with logs. The mill itself was a Little Giant and powered by a six-cylinder Chev motor. It could be set up in a couple of hours. Bill Demaid was the sawyer while Jones tailed the saw and packed the ties away. The species had to be separated—fir, larch and hemlock. Any ties that needed some bark peeled off were thrown to the side and these two would go out after supper to do this. The two Bills worked hard and long hours and sawed between 150 and 200 ties per day, plus side lumber. I believe they received 75 cents a tie.

Our job was to supply the mill with eight-foot-long logs. When the distance to skid the logs got too great the mill would be moved along the road to a new location. We received 55 cents a tie.

Ties now have to have the heart centre in them. Those days they didn't, so as many as six ties could be sawn from one log. The two Bills seemed to get great pleasure by emptying the skidway of logs and then poking fun at us. We got wise after a while and kept some big butt logs close by that we could pull in at such a time to slow them down.
—Walter Anchicoski

Out of the hundreds of applications for FMLs received by the government, several went to larger, well-established Interior companies. The first Interior FML was awarded to Passmore Lumber in 1951 in the Little Slocan Valley. Over the next few years others followed: Western Plywood at Quesnel, Boundary Sawmills in the Kettle Valley, S.M. Simpson at Kelowna, Olinger Lumber at Carmi, Galloway Lumber at Bull River, Cranbrook Sawmills south of Golden, Oliver Sawmills in the south Okanagan, Ponderosa Pine Lumber at Monte Lake, and Clearwater Timber Products in the Cariboo. These were small licences, compared to the Columbia Cellulose FML, and they reflected the existing companies' scale of operations. In 1956, however, the government granted the Celanese Corporation another FML, this one for 2.5 million acres (1 million ha) in the Kootenays, under the name Celgar. As part of the deal, Celgar bought three local companies—Waldie & Sons at Castlegar, Big Bend Lumber at Nakusp, and Columbia River Timber south of Revelstoke—and agreed to build a pulp mill at Castlegar. By this time a scandal had erupted involving the minister of forests, Robert Sommers. He was convicted of accepting bribes for granting a FML to a coastal company, and sent to jail. Only a few more FMLs were issued, after which the government let it be known it would no longer accept applications.

An east Kootenay bush mill, 1930s. FSM 114-9

At the end of the war, returning veterans were provided with financial assistance to attend university, and in BC a substantial number enrolled in forestry studies at the University of British Columbia. They began graduating in the late 1940s and early 1950s, just in time to find employment with the companies that had obtained FMLs: a condition of an FML was that it be managed by a professional forester. The new young forester's immediate task was to undertake an inventory of the licence area and draw up a plan to manage its forests so they would provide a sustained yield of timber for the company's mills. The next task was to convince sceptical mill owners, reluctant bankers and sometimes hostile loggers of the need for self-restraint in the pace and methods of harvesting the company's timber supply. Out of this group of foresters, several became key players in the development of the Interior forest industries over the next three or four decades.

In 1952, two events of major import to the Interior forest sector occurred. The first was the completion of the Pacific Great Eastern railway to Prince George. This was a momentous occasion for the central Interior, if only because construction of the line had begun forty years earlier and, because of delays, had become the longest-running political joke in BC history.

Work began in Squamish in 1912, and it took until 1919 to reach Williams Lake. Two years later the line got to Quesnel, and was stalled there for thirty years. Finally, in 1952, it was pushed through to Prince George, although it took another four years to bridge the gap between Squamish and North Vancouver.

The implications of this rail line for the northern forest industry were enormous. Previously, lumber and other forest products from Prince George and its outlying areas could only be shipped to Prince Rupert for export, or east on the CNR, which meant the industry in this area was oriented mainly to the east, through Edmonton brokers. Now the northern mills could ship through Vancouver, taking advantage of full seaport shipping facilities and the large number of brokers who sold lumber there. Within a few years, mills had begun to congregate along the PGE line. By 1972 almost fifty big sawmills were set up beside its tracks between Fort St. John and Williams Lake, and were shipping 20 percent of the entire province's lumber production.

The second pivotal event of 1952 was the election of the Social Credit government of W.A.C. Bennett. This government was dominated by forward-looking, development-oriented Interior businessmen. There was a natural affinity between Socred leaders and the people who ran the Interior forest industries. After 1956, with the first, unfortunate choice in forest ministers safely stowed away in jail, the second Socred forest minister was appointed. Ray Williston was a Prince George school inspector when he was elected, and he quickly developed a gut-level understanding of the forest industry. He was as honest and fair as his predecessor was crooked. He could sit up all night over a bottle of whisky, hammering out a deal that he and his government would honour. From the day he took on the job and for the next sixteen years, Williston, more than any other individual, shaped and guided the development of the Interior forest industry. By the time he left office in 1972 that industry would be transformed—and practically unrecognizable to those who had worked in it with hand saws and horses after World War II.

In the early days of independent sawmilling the small operators did not have much capital to invest, and what they did have went into trucks, machinery and other equipment. As the mills were portable and the camps temporary, not much money was spent on buildings or facilities.

The usual procedure was to haul several loads of cull lumber from a planer mill out to the new sawmill site. The bulldozer scraped a flat spot down, logs were cut for the foundation, and a floor was nailed on. Two-by-fours were cut to frame the walls and boards nailed on from the outside. Two layers of two-by-twelve were laid across a centre beam, with a layer of tarpaper between them. These were nailed down at the ends of the wall framework to form a sort of boxcar-like roof. A small window opening and a door were cut, and there you were. The only things manufactured were the windows and nails. Even the doors were made of rough lumber. Leather left over from the horses' harness was nailed on to serve as hinges. A large bent nail served as a door latch.

Several of these cabins could be put up in a day. They were usually clustered around the cookshack in a small clearing near the mill. The cookshack was the nerve centre of the mill. It served the usual purposes of cooking and serving food, first aid shack, recreation centre, etc., depending on the mood of the cook. It was also the office, information centre, and in most cases also living quarters for the cook. Everyone came to the cookshack, visitors and crew alike, for gossip and information.

—*Vivian Pederson, who worked as a camp cook in the Houston area.*

CHAPTER 8

Brokers & Bush Mills

The 1950s

The one machine that characterized the Interior forest industry during the 1950s was the portable bush mill. The reasons for this were straightforward. First, the booming North American lumber market was accessible to Interior producers by railway. Second, the supply of fir and spruce was more than adequate, although much of it was in small stands isolated in larger areas of lodgepole pine, for which a market had not yet developed. Third, the kind of large stationary mills built along the East Line and in the southeast before the war were not suited to the condition of the remaining timber supply: logs were getting smaller, and investment in high-speed, double-cutting band saws, for instance, was considered folly.

The response to these circumstances was simple. For a few hundred dollars, anybody with a bit of ambition and no fear of hard work could cobble together a simple sawmill, drag it into a stand of trees bought at a Forest Service auction and start cutting lumber. The only other tools needed were a saw to fall, limb and top trees, and some means—a horse, a Cat or a farm tractor—to skid the logs to the mill. With a couple of friends or neighbours, the wife's brother or her aging uncle, a lumber company could be launched.

The most basic of these mills were homemade. All it took was a hand-hewn log frame on which was mounted a mandrel with a circular saw on one end and a pulley on the other. A carriage of some sort was constructed onto which a log was loaded and run past the saw blade. Adjusting and holding the log in position could be a bit of a problem, but since a lack of dimensional accuracy could be rectified at the nearest planing mill, precision cutting at this stage was not crucial. The lumber had only to be cut to within an inch or so of the desired thickness.

The power supply for the basic bush mill could be just about anything—a belt drive or a power takeoff on a farm tractor, an old automobile or truck engine. If it ran and had more than 50 or 60 horsepower, it could be used to power a small mill. Near Fort St. John, one optimistic lumberman took the engine out of his two-ton truck to power the mill. Once a load of lumber had been cut, he put the engine back in the truck and hauled the load to the planer mill. After a few repetitions of this cycle, he earned enough to buy another well-used power plant, and then it was clear sailing.

When they called them portable sawmills, they were really portable sawmills. They started off pretty small. Most of 'em had gas motors. The Chrysler T120 seemed to be the most popular one, it had lots of quick pickup. Then they started putting big flywheels on these mills so they would carry her right through the cut before it lugged down. My stepfather had a small one out where the Hart Shopping Centre sits. Probably one of the heaviest snow belts close to Prince George. We used to pack gasoline in a ten-gallon drum, on a pole. It seemed like ten miles but I guess it was probably about two miles—a mile and a half, two miles. Then you shovel snow until lunchtime. We had some horses skidding logs then and the snow was up to just about the top of their backs. It wasn't ideal horse logging, so needless to say we didn't make too much money at that one. Yeah, it was the way a lot of small guys got their start.

—*Howard Lloyd. He came to Prince George from Manitoba with his family in 1939, and logged around the area most of his working life. He also served as MLA.*

Foolishly or otherwise, I decided I was going to make a potful of money like all these other guys and I went into the sawmill business with a partner, Jack Milburn. Jack was a classmate of mine. We put together a real portable mill on two big twenty-by-twenty, fifty-foot fir skids and we could drag it through the bush from one place to another, which was fairly typical in those days. It was a good mill of its type. We got a concession in the Willow River. If we could have kept going, we'd have had a quota within a couple of years because that was just when they were beginning to establish the Working Circles. We bust our butt there for about a month (it seemed like years) and that damn strike came along, the IWA strike in '53. It didn't affect us immediately. We were trucking our lumber down to Giscome and loading cars there so we were able to carry on for two or three weeks. Then the railway union decided that they were going to back up the IWA, and this strike, which surely wouldn't last more than two or three weeks, lasted a way on 'til the end of January. When it was over we ploughed out. God knows Willow River snow is pretty deep. We ploughed out, but the lumber market was down $20, as I recall, and our banker had just about had it. The cost of ploughing out and getting going again in midwinter was...anyway, we went belly up.

—*Bill Batten. He graduated in forestry from the University of BC in 1951, and worked for Evans Forest Products at Golden through most of the 1970s and 1980s. He now works as a consultant and manages a Woodlot Licence.*

My family came to Vanderhoof from Saskatchewan in 1942, during the war. I think the first few years were kinda tough but they survived and got started in a sawmill. Dad brought three horses with him from Saskatchewan. He started with the three horses and he used two out in the woods. My dad didn't actually get involved in the woods that much, it was my older brother. He bought a $25 power unit, an old Chrysler car motor and a $500 sawmill. And with the two horses he skidded logs. The first winter, this was probably about 1945 I believe, we cut all winter by hand, felled 'em, bucked 'em up and skidded 'em into a landing and decked 'em up for spring. And in the spring we bought the sawmill. It took us all summer to cut them. Prince George had some mills that bought rough lumber and they processed it from there. Did that for a few years and then my dad and my brother, they bought a small planer mill. Set it up right here in Vanderhoof. It was all bush at that time. You just hack out enough by hand to set up the planer, and start planing lumber. BC Spruce was the name of one outfit that bought most of the lumber. If I remember, the first carload of lumber we ever loaded, it took us probably about three or four days to pack it into the boxcar. We had dumped it along the siding and we packed in every board, probably a couple of hundred feet. Packed it in a board at a time. That was a big deal, you know, shipping out a boxcar load of lumber.

—*John Martens. He and his family built Plateau Mills at Vanderhoof, which was taken over by the provincial government in 1972.*

Page 146: *A portable Holding Lumber mill, probably used to cut bridge timbers and boards for a plank road.* BCARS 73903

Page 147: *Filing the saw at a bush mill near Enderby, 1952.* EDMS 2708

In other cases, bush mills were a stationary-mill owner's solution to a lack of capital. For example, by the time Bert Balison and his two sons had moved to Heffley Lake from Vancouver in 1946 and bought a small stationary mill, complete with a bunkhouse, a cookhouse, a team of horses and 350,000 board feet of timber, their capital was gone. They cut

A portable bush mill and edger, supplied by logs skidded in with horses. There were hundreds of such operations throughout the Interior in the 1950s.
BCARS D-05565

logs from the timber stands around Heffley Lake, skidded them into the water and used an old outboard to herd them to the mill. When the skidding distance to the lake became too great, they had a problem. They could not afford a Cat, nor could any other loggers in the area. They dismantled their stationary mill, reassembled it on some log skids and dragged it into the woods with a winch powered by a Model A Ford. When that worked, they built a couple more mills, each with a four- or five-man crew.

Now that they were in the big leagues, they could afford a Cat to drag the mills around, as well as to skid logs. Then they got a truck to haul rough milled lumber to the stationary mill site where they had installed a planer. With this machine, they quickly found, real money could be made. They started buying rough lumber from other bush mill operators. By 1951 they had closed their own bush mills and reassembled the stationary mill on the railway line. That year they adopted a fancy new name, Balco Forest Products. Throughout the 1950s, they

Trucking rough lumber from a bush mill to the Ketcham brothers' recently purchased Two Mile Planing Mills Ltd., Quesnel, 1956. WFT

and their forty employees put out about 40,000 board feet a day. From these modest beginnings, the Balisons' operation evolved into a major Interior company.

> I come up to this country in 1954 and worked for a gyppo contractor out of Fort St. John. Probably thirty, forty thousand a day is what they sawed, most of the mills. And there was smaller ones than that too. Depending on the crews they had. Not closed in, just something to keep the snow off the head rig and an edger behind it and that was it. You piled the lumber on blocks and the truck backed into it and the blocks fell over and fell on the truck. You put a chain around it and hauled to either Fort St. John or Dawson Creek where the planer was.
>
> The mill was set right up in the berth. You skidded the logs in with Cats or sometimes tail-dragged them behind a truck, two or three logs at a time. You bucked it right there at the mill and had horses skidding the logs onto the skidway. A lot of labour involved in those days. It kept you warm that way.
>
> I started out as a tail sawyer. I came three winters before I moved to BC and stayed. I went back to Saskatchewan and worked on road construction in the summertime.
> —*Jack Thompson, who retired from Canadian Forest Products Chetwynd Division in 1997.*
>
> I graduated out of high school in 1948 and worked in some of the local sawmills for a while after that, J. Young & Sons was one. Learned how to operate a D4 Cat. It was a real big piece of equipment in those days. Lots of plumbing and rigging on them. The tracks were so wore out if you went over a log on a sidehill, you'd throw at least one track off and sometimes two. And they were working in really hilly ground off the river there, so we had lots of fun learning how to operate them. Before they got the Cat they used horses to skid the logs into the sawmill. As the bulldozers first came in, they would use them for pushing roads, then they'd put the rollways out and skid the logs out of the bush with the horses, roll 'em down the rollway and either load 'em with an A-frame or just load 'em by hand. And we done that at the little mill up the river that I worked at. We logged with horses 'til it got a little too far, then they tried rubber-tired tractors. And rubber-tired tractors, after about two weeks on the same road, you couldn't get 'em back to the bush empty. They didn't work too good, so then they went to Cats. In the wintertime they'd usually pull 'em down tree length and in the summertime they'd load them on the sloops. If you got a helper it was a little easier, and if you didn't, you loaded 'em by yourself.
> —*Howard Lloyd*

Not all bush mills were primitive devices. For decades it had been possible to buy small, portable mills, and lots of them had been put to work in various parts of the Interior and across the northern prairie provinces. Scores of manufacturers across the continent produced them and they came in a wide range of styles and configurations. By the early 1950s a number of highly portable models with steel frames were available, some of them mounted on wheels. The Ainsworth brothers, who pioneered the large-scale marketing of lodgepole pine in the Interior, bought this type of mill when they decided to get into the lumber business in 1952. They bought a Jackson Lumber Harvester, built in Jackson, Mississippi, which could be towed behind a pickup truck and set up for business in less than an hour. In a single day, just to prove they could do it, the Ainsworths cut lumber at three separate sites.

The Ainsworth brothers, en route to the Cariboo with their portable mill, 1952.
Ainsworth collection

When I first came to the Interior, there were a lot of portable mills still cutting out in the bush and we were buying their lumber. That was a very interesting education for me, to be working with portable mills, because they could be working 'til Friday night and leave a bunch of fuel bills and grocery bills and by the weekend they were gone up to the Cariboo. Then we would get the fallout from that, phone calls from grocery stores and fuel outlets. "What happened? We haven't been paid!" It was a real mess. Some of them were cutting Federated Co-op timber and then taking the lumber into the sawmill. And a lot of them had their own little sales that they operated on and brought the lumber in for sale and some of them were real haywire operators. It was off-size. It was a real headache to try to administer. Then when the quota system came in in '57, '58 and '59, your average over that three-year period became your quota. And that set up the basis for the sustained yield aim of the provincial government. Looking back, that brought a lot of stability to the industry and made it possible to put in stationary mills. I personally think it was a very good move in the long run for the government and the industry and everybody else concerned, especially the taxpayers, because it cleaned up a real haywire mess.

There was opposition to it, no doubt about that. Some of the smaller portable operators said, these big outfits are gonna take over and we're out. There was some of that, but it didn't receive the same amount of attention that it received on the Coast, that's for sure. As I say, I think it was a good move. Brought in a lot of new investment. And as that happened, and the timber all came to big central mills, you got the opportunity to co-ordinate the chip supply.

—*Fred Stinson. He grew up around logging camps in the Fraser Valley and, after serving in World War II, graduated in forestry from the University of BC. He went to work for Federated Co-operatives at Canoe in 1957, and in 1967 began to work for Northwood Pulp & Timber Ltd.*

A lot of bush mills were brought into the Cariboo and northern BC just after World War II by refugees from Saskatchewan. By the beginning of the war, a substantial sawmilling industry built around bush mills had grown up throughout northern Saskatchewan. After the Co-operative Commonwealth Federation (CCF) government of Tommy Douglas was elected in 1944, it set up a lumber marketing board to which all producers were forced to sell lumber. Since they also had to buy their timber from the government, they were caught in what they considered an intolerable situation, in effect working for a government most of them had voted against. The booming lumber market in BC was a tempting alternative, so almost all of them packed up their mills, loaded them onto CNR rail cars and headed for Prince George. Unlike many refugees, they arrived with well-developed skills and the primary tool of their trade—a portable sawmill. Their timing was right and they flourished in the booming market. No one ever counted how many Saskatchewan lumbermen made the trek to BC, but there were likely two or three hundred of them working around Prince George and through the Cariboo.

A small Okanagan bush mill, 1950s. KLM 12560

Cliff Falloon was one of them. He came to Quesnel from Saskatchewan in 1948 and did contract logging for Oliver Brownmiller, who was in the process of building up a mill at Quesnel. Falloon brought a Cat and a tractor with him and, after logging for a couple of years, traded the tractor to a farmer for a portable mill. He needed an engine for it and Brownmiller had a six-cylinder Chrysler T120 he didn't need, so they went into partnership and set up B&F Sawmills. After Brownmiller sold his mill to West Fraser, he got out of the partnership, which then became known as Falloon Brothers Sawmill. It stayed in business until it burned in 1969. Falloon went back to logging with his sons under the name Quadra Logging, which in 1998 was one of the larger full-phase logging contractors in the Cariboo.

There was no set pattern in the operation of bush mills. Sometimes two or three guys would drive out to work in an old two- or three-ton truck, and spend the morning logging and the afternoon cutting lumber. They'd load up the truck and drop their day's production off at the planer mill on their way home. At $30 a thousand board feet, they would each take home $10—if they had a good day.

> Although I spent most of my years in the logging, I started driving truck hauling lumber from all these different mills. Loaded it all by hand. Some of the mills, they'd have put their lumber in the most awkward position you could get. When you had a big pile of lumber it wasn't too bad, but the load got higher than the lumber and the deck pile got lower. I used to have to set the lumber up on the edge of the truck and then crawl up the truck and pull it up. I did it all by myself. Same in the boxcar. Used to pack it all in by myself. Sometimes there was people in town, some young guys that weren't working I could get ahold of and hire, I got them to help finish the top part, that was a little difficult. But if there was no one around I just packed it in. Fill it right up, tight as you can.
> —John Martens

Right: Falloon Brothers' arch truck with a turn of logs at Cottonwood, 1966. WFT

Below: Pete Byspalko loading logs at a Brownmiller logging site for A.S. King Logging, with a forklift mounted on the rear of a farm tractor, 1956. WFT

Some of these operations were financed by lumber brokers from Edmonton, Vancouver or the USA. The Robert E. Malkin Lumber Company in Vancouver bought and sold lumber, set up portable mills throughout the Interior and financed independent bush mill operators. It also built and sold its own portable mill, the Bach Mobile Sawmill, which the company claimed could cut 20,000 feet a day with six men working on it. It combined in a single unit a head rig, carriage, edger, power plant, slab conveyer and sawdust blower. "In case of fire," Malkin advertised, "you just pull your mill out with a Cat or truck in a minute." Malkin eventually went broke trying to develop a portable band mill at Revelstoke, leaving Sidney Eger, one of its young immigrant employees, out of a job. But there were a lot of people willing to bridge the gap between the bush mills and the international lumber markets, and before long Eger went to work for another Vancouver broker, Leslie Kerr.

Kerr was a charter member of a group central to the development of the BC forest industry during and after the war—European refugees with an intimate knowledge of global lumber markets and connections to sources of finance. He was from Yugoslavia, where his family had worked in the lumber trade for three generations. He received his doctorate in business at the University of Vienna, and when war broke out he was working in Poland for the Koerner family, who played a central role in the Czechoslovakian forest industry and who fled to Canada ahead of the Nazis. They set up a coastal company in Vancouver, Alaska Pine, and brought a lot of forest industry expertise along with them.

> My brothers had worked on the carriage at Alexander's and up the river sawing and dogging, and we'd worked in different sawmills around the area. There seemed to be quite a demand for lumber after the war, so my brothers and I bought a sawmill out in the Sweet Creek area, about twenty miles west of Prince George. A small, portable sawmill. Morris Sims had it there. We were about a mile and a half off the road and it was a terrible road. We didn't have gravel trucks in those days, so it was a little tough getting the lumber out in the summertime if the weather wasn't co-operating. And like most small ventures, you know, you work at it for two or three years before you take any wages out of it, but we'd take a drag each one of us when we really required it. Big thing was to try to keep the sawmill running. We had one pickup between the four brothers, so it was a pretty important rig. And the odd time when you got it in the garage the dealer'd say, "We got you this time, boys, you're gonna pay your bill before you take that pickup out." And we'd say, well, keep the pickup and we're out of business, so suit yourself, you'll never get paid! "Well, I guess you'd better take it!" Anyway, we stayed at it. We moved it a couple of times. It was on skids for a while and we did a couple of settings rather than truck timber to it. And then we moved the entire operation to Cluculz Lake and set up a little bigger mill up there.
> —Howard Lloyd

Another prominent Czechoslovakian family that came during the same period was the Bentley family, which founded Canadian Forest Products. One of their associates was a brilliant, cultivated young engineer named John Bene, a Hungarian whose family had been in the plywood manufacturing business since 1834. After helping the Bentley group establish Pacific Veneer in Vancouver, he set up his own company, Western Plywood, in 1945. Most of Bene's timber supply came from the Cariboo, and in 1951 he built a plywood plant in Quesnel to make spruce and fir plywood. Ten years later he joined forces with three other firms to create Weldwood of Canada Ltd., with himself as president.

When Bene started buying timber he ended up with a lot of sawlogs not suitable for veneer, and in 1946 Leslie Kerr set up a sawmill in Quesnel to produce finished lumber from these

A.S. King Logging's crawler-mounted loader working at Bellos Lake in the Cariboo, late 1950s. WFT

logs, which he then marketed through his own Vancouver sales firm, Lignum Ltd. Two years later he built another sawmill and planing operation in Williams Lake and began buying unfinished lumber from bush mills in the area. Over the course of the next decade or so he forged a relationship with these mills that was prototypical of this era.

When Kerr began, the lumber from most bush mills was dimensionally inconsistent. A board might be an inch thick at one end and two inches thick at the other. Planing this stuff was costly. Kerr encouraged the bush mill operators to improve their equipment and milling techniques, providing them with interest-free loans they paid off by delivering lumber to him. In some cases they cut logs they had obtained at Forest Service auctions; in others they cut Kerr's timber. What evolved was a flexible arrangement between Lignum, which finished and marketed the lumber abroad, and the bush mill owners, who logged the timber and produced the lumber. Lignum ran its own sawmill to keep a handle on costs and techniques.

When Kerr set up shop in Williams Lake, he was the first in town with a planer, and there were about 25 bush mills operating around the area. By 1956 there were a half dozen planer mills and more than 150 bush mills within the operating area of Williams Lake. This pattern was being repeated throughout much of the BC Interior at this time.

> I love the life I've had. I did some good logging for Penny. We developed the Slim Creek area—winter logging, decking it on the banks of the Fraser and river driving during the summer. I love the Fraser River. It was easy driving the Fraser, but the danger was you'd lose your logs. If they went past your mill, you've had it, Mac. Sometimes they did, too, particularly in the early spring. If you got your logs in too early, before the river was high enough, and then the river came up, you could get logjams and then when they broke you lost everything. So, yes, it's easy in the sense that there was lots of water, but we were driving down some of the side creeks too, the Torpy and Slim. I drove Slim one summer. It wasn't very successful. You had to catch the big runoff in the small creeks. They would never let you do those things now, of course.
>
> I got to know the whole river very well, from McBride to Shelley, between working for different people we got our logs from here, there and everywhere. If you were at Penny, you got your logs upwater from Penny. When you were in Shelley—we used to say that Shelley got everybody's logs 'cause they were about the last mill down the line. Anybody lost a log, it was at Shelley. But you know, even in those days every once in a while you'd get a report someone found your timber mark down at the mouth of the Fraser River.
> —*Bill Batten*

Another early arrival in Williams Lake was Gabe Pinette, along with two members of his wife's family, Dollard and Robert Therrien. Pinette had grown up on the prairies, run bush mills in Ontario and worked at several coastal BC mills before he and the Therriens headed out on their own to Williams Lake. They started with a bush mill a few miles outside Williams Lake in 1953: by 1955 they had a sawmill and planer mill in town and were finishing lumber for themselves and others, under the name Pinette and Therrien Planer Mills. In the early 1960s this company, along with the Ainsworth family's company, pioneered the use of lodgepole pine for lumber.

In a slightly different manner another family set up in the Cariboo at this time. The Ketcham brothers—Bill, Pete and Sam—grew up in Seattle, working in their father's small lumber wholesaling company. In 1955, thinking it was time to strike out on their own, they got in a car and headed up the Cariboo highway looking for an opportunity. At Quesnel they found a small planing mill, Two Mile Planing Mills, for sale. They bought it and Sam stayed in Quesnel to run the show, while Bill and Pete went back to Seattle to look after the business and sales end of the enterprise.

Following pages:
The Ketcham brothers' first mill, Two Mile Planing Mills, Quesnel, late 1950s. WFT

> When I came [to Chetwynd] there probably was, I would think, something like forty guys working at the mill. It was owned by Fort St. John Lumber and the bunkhouse was in the annex of the hotel, so they kind of kept the money circulating. The guy that owned the sawmill owned the hotel too. So he kind of took it out of one hand, put it back in the other. It was the only reason I could see that he wanted it within walking distance of the hotel, because he built this frigging sawmill here in the muskeg and we been fighting it ever since.
> —Jack Thompson

West Fraser's first sawmill, at Bald Mountain, 1959.
WFT

Sam started buying up bush mills around Quesnel (he became known as "Bushmill Sam" in the Cariboo) to feed the planer, and within a few years had set up another planer in Williams Lake. The timber for this mill came from the west side of the Fraser River; believing mistakenly that this would become the centre of their operations, the Ketchams named their company West Fraser Timber.

The development process taking place in the Cariboo in the early 1950s was also unfolding elsewhere, particularly in Prince George. Along the upper Fraser, many of the East Line mills were in a state of decline, having consumed the timber stands for which they were built. As was the case in the rest of the Cariboo, the country around Prince George was colonized with bush mills. The first planing mill in the city had been built in 1939 by Martin Caine. He was an Englishman who joined the Klondike gold rush in 1898 and, after spending the war in France, got a contract from the Grand Trunk Pacific to cut ties on the East Line. He built a tie and lumber mill there and ran it for almost twenty years before relocating in Prince George, just in time for the postwar boom.

> Summer ground was always a problem. We didn't have any Section 88 or the Forest Service building all-weather roads for us, so if the market wasn't strong we really couldn't afford to do all that much gravelling on haul roads, particularly in this country up here where you had fairly low volume. We used a LeTourneau for a while but it was only a two-wheel drive. Bigger ones they brought out later with the four-wheel drive probably would have been a little more adaptable. They weren't a high-speed machine going back to the bush and they were pretty blasted heavy on the roads. They'd pull an awful load but they're awful hard to control coming down hills. We had a steep hill at Cluculz and it had a couple of hairpin turns coming down to the sawmill. Every now and again the trees would jillpoke around this arch and drive it right off in the bush instead of making the corner. So my brother Ben got on the Cat one weekend and built it straight down, just eliminated all the turns—but man, was that steep. When you went over the top, as soon as the trees cut loose you got one surge that got rid of all the traction and you'd skid the rest of the way to the bottom of the hill. Kind of a wild way of getting the timber down.
> —Howard Lloyd

By 1955 there were a dozen planing mills in Prince George, lined up along the Nechako River on what was called Planer Row. Throughout this period they were supplied by as many as five hundred portable mills working in a large area surrounding the city. Between the war's end and 1955 the population of Prince George grew from three to twelve thousand, almost all of this growth fuelled by expansion of the forest industries. Compared to the evolution taking place in Williams Lake, the transformation of the industry in Prince George was quick, free-wheeling, almost chaotic. Practically anybody could get a timber sale and haywire a bush mill together, and at times it seemed as if every man, woman and child in the city was doing just that. Most of the forest land around the city had been designated as a "special sale area," with the idea that it would eventually become farmland, so the sustained yield restrictions on timber harvesting which applied elsewhere after 1948 were not in force here.

Not every new operation established in this country in the decade or so after the war was of the bush mill variety. A few larger, more stable ventures were launched. One was at Hixon, midway between Prince George and Quesnel. In 1950, as part of the Saskatchewan lumbermen's migration, the Dunkleys loaded their planer on a rail car, picked up a new sawmill at Prince Albert, and headed west. They bought a piece of farmland right along the soon-to-be-built Pacific Great Eastern Railway and set up their mill. They began logging directly to the mill with horses, and when the nearby timber ran out they switched to Cats and arch trucks. The logging was done in the winter by Saskatchewan loggers, who returned home to farm in the summer. At first the Dunkley mill trucked its lumber to the railway at Prince George, but within a couple of years the PGE came through and they began shipping to Vancouver and selling through brokers there. In 1965 Dunkley and two other Interior companies formed Overseas Spruce Sales and took out the first Interior membership in Seaboard Lumber Sales, a co-operative marketing agency set up in 1935 by a group of coastal sawmills.

Loading logs by hand at Art Holding's logging operation in S.M. Simpson timber near Kelowna, early 1940s. Logs were hauled to the skidway with horses, then rolled onto the truck with peaveys. Left to right: Roy Brown, Jack Riviere, Bruce Hallaln, Halley Fowler. BCARS 73933

> When I was falling for Kootenay Forest Products, it was pretty well the year round. The biggest layoff KFP had was in the spring breakup season. When I supervised, I worked all winter and I worked all summer. The only time I had off was breakup time, when it was mud up to your knees. That's when you had your holiday.
>
> If you wanted to fall all winter, you could. I fell in ten, twelve feet of snow, with snowshoes. I'm telling you, it was dangerous work. If there was ten feet of snow, you shovelled down six or seven feet around the tree. You'd tramp it all down so you were within maybe a two-foot stump. You could just see out of the hole. You're up to your neck in snow. Then you had to make yourself a ladder to get out of there because when the tree starts to go, you had to get out.
>
> You made two steps, and packed them well with the flat side of your axe, then you put your snowshoes on them and when the tree started to go, you just jumped from one to the other and you're on top and you're free. If the step breaks, then you're right in the bite. There's been the odd man hurt that way. Once the tree is gone, then you pick up your snowshoes, put them on and go to the next tree.
>
> Carlos Waldie, the KFP woods superintendent, got the brain wave that they would try high-lead in this country, so they got a spar tree and a riggin' crew together and they went up Grohman Creek. I cut the timber up there for about six weeks, way up at the far end of the road—I called that the last log that was up there. I cut the first one for the right of way, and I cut the last one for the high-lead. But the high-lead didn't pan out in this country too well because there's not enough concentration of timber. Too scattered.
>
> Later KFP tried two steel spars, the big ones where you'd drive them along the road, put down the pads and put down the guylines, but they never paid off.
>
> The only cable system that would pay off in this country was the jammer. It was easily moved and fast to set up, and it worked on what they call a slack line. They had two drums on a jammer and they could go down in some bad draws, especially on the lower side of the road, and bring up logs. They would take a strip forty or fifty feet wide and when all the logs were picked up out of that strip, they moved another forty or fifty or sixty feet to another stump and then cleaned that up. They could log some very bad ground fairly cheaply with jammers, but there was no money in high lead.
>
> —Oscar Schmidt. He was born in Alberta in 1915, and worked as a faller and bullbucker on the BC coast and in the Interior for his whole working life.

Dunkley Lumber evolved into a unique operation. It acquired an excellent timber supply in the heavily timbered valleys east of Hixon and did all its own logging. The company was a closely run family enterprise with a fiercely loyal group of workers, whose families created a community around the mill site. When the Dunkley family decided to sell in the 1980s, it could have sold the whole business to any one of a dozen buyers. But instead of selling out to the highest bidder, who no doubt would have closed the mill and trucked logs into Prince George or Quesnel, they decided to sell to a family-run company.

After much searching, they met the Novaks. This family, like the Dunkleys, consisted of six brothers and two sisters. One by one, beginning with the eldest, Tony, they had escaped through the Iron Curtain from Slovenia. In Canada they settled around Quesnel and Prince George and ended up logging, starting with a single skidder. Tony was involved in a small sawmilling operation at Quesnel for a while, but he concentrated on logging and had little to do with running the mill.

The Dunkleys wanted to sell their company to the Novaks. They went to some trouble to persuade a bank to lend a substantial sum to this immigrant family with almost no sawmill experience. A couple of the Dunkleys stayed on for a few years to keep the operation running smoothly. The Novaks were enormously successful and poured their earnings back into the mill. By 1995 the Hixon sawmill had become the most technologically sophisticated sawmill in BC, if not in North America.

A Model 30 Caterpillar with a well-loaded sloop of logs, near Adams Lake, 1940s.
BCARS 25514

After they moved the mill into town, then I started to go to the woods logging. Spent most of my years in the logging after that. We had a 1950 TD9. Great big. That's what we used to build roads and do the skidding. When we started hauling logs in 1955, we got another TD9 as well, but it had a front-end loader and a winch, and you could change the fork into a blade. You could build roads and landings and then skid with it and load logs in between. You don't produce an awful lot that way but it was enough. Once we started using the Cats then we didn't use horses. Except one winter there was a fella here that thought he could still skid with horses and compete. He tried it and he lasted about two months, then gave it up. That's the last time I ever used horses. I started skidding horses when we first started. That's a job I said I'd never do again unless my wife and I were starving. That's the only way. I felt sorry for the horses. It's not a good job for horses. Once in a while they'll knock down a dry snag and it'll fall on top of 'em. I very much disliked that job.
—John Martens

Another combined planer and sawmill that started from scratch opened in Horsefly in 1955. Harold Jacobson and his brother O.J. had been familiar with the business for a long time. They had put in a few years logging on the Coast and later a few more years in the Okanagan. Their father had started horse logging in 1927, in Terrace, and had built a small water-powered sawmill and spent most of his life in the forest industry. In the mid-1950s the

We set up a mile out of Horsefly, in '54. In '55 we started operations. It was a good permanent mill. It wasn't a high production mill, about a million feet a month. We were probably the second mill that didn't go along with the idea of cutting for a planer mill. We put our own planer in. And there was another mill that had a planer. We did our own marketing right from the word go. And that was one big change that came about. There was probably a hundred or more small operators supplying rough wood to planer mills here in Williams Lake, and they began to phase out. One after the other they began to fall by the wayside and we came in with a complete stationary sawmill, hauled our logs in. And the planer mill. And did our own marketing and all the rest. Well, from then on the little gyppos had a rough time, and they just disappeared. Some we bought out, some went broke, some just pulled out. It ended up there was only really five or six operations in the Williams Lake area. Just through simple, natural whatever it takes to stay in the race.

—*Harold Jacobson, one of the founders of Jacobson Brothers Forest Products in Williams Lake.*

[After I got my own logging company] we bought our own arch trucks. We started off with the small gas jobs. They'd move a lot of timber, those little trucks. Really tough and really get around. Probably about a five-ton tandem. It was just amazing the amount of wood those things could move, as long as you didn't have any steep grades to pull. The ones coming downhill didn't matter too much, the truck was fast enough it would outrun the logs. As long as nobody was trying to come up the hill you were all right. That happened the odd time too. We had radios on the trucks pretty well from the first days there. They were supposed to wait for a truck or wait for an "all clear" before they headed up to the bush. There was a few times they didn't. I'll have to say a few times I didn't, but after you have a couple of close escapes you remember the next time. You can usually miss the truck all right, but the logs would fan out so much they would get you. It wouldn't be too bad if they just wiped you off the road, but if it was on a corner they'd come right over top of your truck and that wasn't very nice.

If there was no adverse grades, they could handle a couple of Cat drags fairly easily, thirty or forty logs. As we moved into the diesels, it was just amazing what they would pull. We had a company operation that I worked on before where we were logging to Summit Lake, and they built a private road on the far side of the Crooked River all the way down to Davey Lake. We used our small truck in the bush and brought it down to a real big landing. They would bring their big Kenworth from the sawmill and pick 'em up from there. And at times it just looked like that whole landing would come alive. They would have fifty or sixty logs behind that thing, for twenty miles. All on ice and snow in the wintertime. We built fairly wide roads. It was probably harder on the wildlife than on the logs there. If you got really deep snow and a moose got on the road they were pretty reluctant to leave. The truckers'd follow 'em and follow 'em and follow 'em and finally the moose would get up on the snowbank and then the trucker could get his truck by. They would try to give 'em time to get out of the way but quite often they'd kind of rub 'em on the legs and try to push 'em over top of the bank. But it was just amazing the amount of timber they could move on those things.

We had a final drive going on a D7 Cat out in the bush, so how are we gonna get it in? He was just hooking up a load of trees on the arch truck and I said, I wonder if you could pull this Cat in if I just ran it on the back of your logs. I was kind of joking him. "Why don't we try?" he says. "Hook her on. There's no bad banks between here and the camp. If it doesn't go over a bank someplace, maybe it'll pull 'er in there." So we put the D7 up crossways on the logs and he give 'er a snap and a snort and away he went and just dragged her to within about a quarter of a mile from our shop. We limped it into the shop right from there. After that if we needed a Cat in camp, just load it on the back of the logs and he'd haul that sucker all the way in.

—*Howard Lloyd*

Jacobsons decided to avoid the bush mill route, so they put up a stationary sawmill and planer at Horsefly to cut fir and sell it themselves through the Vancouver market.

At that time there were a hundred or more bush mills working in the surrounding area, and Jacobson Brothers Forest Products began buying from them as well as finishing their own lumber. After nine years, with the bush mill era in decline, they moved to Williams Lake and set up a state-of-the-art band saw mill. As the fir stands east of Williams Lake were cut out, they began to move into lodgepole pine west of the Fraser River. While most of the other mills in the Williams Lake–Quesnel area moved toward high-production dimensional and stud mills, the Jacobsons concentrated on improved utilization and efficiency in the production of high-value pine lumber. In 1994, they sold their sawmill to Riverside.

At any given time through much of the 1950s, there were as many as two thousand bush mills operating in the Interior and supplying lumber to a few dozen planer mills, and an equal number of combined sawmill and planer operations. It was almost impossible for the Forest Service to keep track of these mills as they moved from one timber sale to the next, from one contract to another. There were always some who, disinclined to pay their stumpage or to pay

P.F. Liebscher's mill at Cedar Creek, near Cranbrook, 1955. The mill was built by Cranbrook Foundry. FSM FS65.03

log suppliers or other creditors, simply delivered the last of their lumber to a planing mill at the end of the week, pulled up stakes and disappeared. By Monday morning they were set up in another district, perhaps under a different name.

Most mill operators, however, sought a higher degree of stability. As the number of mills grew, competition for timber intensified. In the old days, when there was room for everyone in most regions and someone applied to the Forest Service for a stand of timber, it was rare that anyone bid against them. As more mills appeared and the better stands of trees disappeared, timber auctions became livelier affairs, with substantial bonus bids offered.

Under these conditions, it was inevitable that a whole host of sharp practices should evolve. A newcomer to an area could find himself involved in spirited bidding against established operators who sometimes escalated prices to unrealistically high levels and then stopped bidding, leaving the greenhorn in possession of timber bearing stumpage fees too high to pay. Less scrupulous bidders who were not actually interested in buying timber could withdraw from the auction in exchange for cash or some kind of deal on a logging contract or another stand of timber. If two genuine bidders appeared at an auction it was common for them to call for a recess, leave the room and cut a deal in which one of them would get the timber at a low rate and the two of them would jointly work the sale. There were dozens of such ruses and tricks. In some areas the competition became heavy when the amount of timber logged each year reached or exceeded the annual cutting limits determined by sustained yield policies adopted in 1948. In these cases the demand for timber exceeded the supply, and even established operators had to compete vigorously for logs.

> We used to just put up a block. Where we seen some timber we wanted, we'd mark the area out, go and tell the Forestry. Forestry'd come out and bore it to make sure it's the right age to be cut. It was selective logging. Most blocks was eleven inch or nine inch, anything smaller than that had to stay. You selective logged and if you bruise the trees up, mark the trees up or if you had a stump that was undersized, you'd better have a good explanation or you were charged. It worked great. You just farmed the timber basically. It was a good system. It really worked good.
>
> You went out and you marked out a block of timber, close to your farm is what normally was the thing, because we had horses to take care of too, eh, so you want it fairly close. They'd put it up for auction. It would come up but no one would even dream of bidding against you. It was a gentlemen's agreement. I mean if it was close to your place no one would even think—if you put it up it was basically your timber. There might have been the odd case that a person broke the rule but I don't think he lasted too long in the area if he did. It was pretty cut and dried. So everybody worked and everybody helped everybody. If somebody had a block of timber and the other guy was waiting for his to get approved, they'd split the block of wood and then split the next block, they'd move their mill in on it too. It worked really super good. Everyone worked.
>
> When big business came in, she wasn't business as usual at all. And then, of course, the government worked hand in hand with the big mills and they figured it was easier for them. They get the volume out, less headaches 'cause they only had to deal with a few parties. Pretty soon the little fellow was forced to get out. We had no more quotas. You couldn't buy timber. Not many years back, even if you tried to get timber for a small mill, they made the regulations pretty well so you couldn't run a small mill any more. Even if you could meet all the specifications you didn't have a chance, because the mills had bidders at any price that would go in. So when the sale went up, if you went in for a sale, they would hire somebody to bid the sales right out of reason. But it didn't matter to them because it was small volume. When it averaged out over their overall amount it was very, very minimal.
>
> When quota came in we got nothing. They give it all to the mills. They just squared off the whole country in about three sections, and the mills fought among themselves in order to get the bigger chunk of the pie. The local people were forgotten.
>
> —*Bruce Kerr*

This period in Interior logging history is described in one of two ways. One version depicts it as the era of true free enterprise, when competition was real and timber prices were determined by the elemental forces of supply and demand. Anyone could get into the business by paying a higher price for timber, and the less capable were forced out. The other version is that this period was one of chaos and uncertainty, which discouraged investment and development of the industry.

> It was not only the coastal industry didn't want to pay any attention to the Interior but the Forest Service didn't want to pay any attention. Or at least the chief forester didn't want to pay any attention to it and so, if you go back through the records, you'll find that Dr. Orchard, when he was deputy minister and so on, he never believed there should be any basic reforestation in the interior of the province. He never believed there should be any and what there was, in a very limited way, was confined to the coast. But the major companies, the five major companies—in my sixteen years of administration, I never felt I could get anywhere with them. The five major companies at that time seemed to adopt the attitude that they knew more than anybody else and even took the stand at certain times, if you go back through some of the hearings, that when there should be new policies or new distributions of timber, I should go to them and ask them how that should be done. They were very, very set in their ways. They never did move and keep up with the times. And I never got one imaginative suggestion out of the coastal operators anyway.
> —Ray Williston, BC minister of forests from 1956 to 1972.

In any case, it came to an end in 1960 when Ray Williston directed the Forest Service to set up a quota system. This system gave "established operators" a timber quota based upon the amount they had logged over the previous three years. They received bidding privileges on this volume of timber at auctions which effectively guaranteed them a specified volume of timber each year and weakened the position of competitors at auctions. Williston, who preferred to bring about change by introducing new regulations instead of changing laws, never did provide the quota system with legal standing, and it stayed in place as the keystone of the province's timber allocation system until a new Forest Act was passed in 1978.

A primary effect of the quota system was to make possible the acquisition and concentration of logging rights. Shortly after operators' quota levels were established, it became possible to buy and sell quota. Hundreds and hundreds of bush mill owners were assigned quota when the system was set up, and most of them soon sold their rights to the planer mill operators with whom they worked. Through one means or another, cutting rights began to accumulate in larger volumes held by fewer companies.

The industry on the Coast was undergoing a similar process. Initially a significant portion of quota was held by a large number of independent market loggers, and the coastal log-transportation system enabled them to sell their logs anywhere on the Coast. The battle over quota in this region was intense and bitter, with most market loggers, and the independent mills they supplied, being squeezed out of the business during the 1960s and 1970s.

In the Interior, logs could not easily be moved over long distances. There were few market loggers and no log market. Most small sawmills—the next level in the processing cycle—had well-established working relationships with one or more planing mills. An operator might not be particularly happy with these arrangements, but they were in place and not easily changed. In most cases, quota passed from bush mill operators to planer mills voluntarily, without acrimony or conflict. Williston viewed the money received from quota sales, for which the licensees had paid nothing, as the operators' reward for having built up the industry.

In the Prince George District, the quota holders in each Public Sustained Yield Unit or Public Working Circle (the geographical areas within which allowable annual cuts and quotas

> I screwed Dave Ainsworth up twice. In 1956 or '57 I thought we were overcutting in the Cariboo so I naively—I was pretty naive in my day—naively went in and said to industry, Just hold your cut right now. Just hold your cut because we're gonna take an inventory of this and see what it is. So hold your cut for the next year or so, so I don't get into too much trouble. So in the next year or two, everybody in that 100 Mile area had doubled their cut except Dave Ainsworth, and Dave Ainsworth stayed exactly where he was. When I finally got the inventory straightened out I had to cut everybody 54 percent. When I did that, pretty near every one of them came back to where they were when I said stop. But poor ol' Dave, he lost 54 percent by standing still. I really goosed him! So then Dave came to me and said, "Nobody's cutting any of this pine around here. I'm gonna take a crack at it so put up a pine sale for me and I'll see if I can do it." I put up a pine sale and as soon as I put it up—we didn't have it in quotas—some of the bigger guys took a look at this and thought, Oh gosh, we can't let this guy get started on pine. We're gonna have to bid that up. Well, Dave didn't have any money to bid against these other fellas so he went to them and said, "Look, if you want to take the pine sale, take the pine sale, but I'll make a commitment to you. I'll cut all the pine, I'll do all the business and you don't have to worry anything about it. If you want the pine sale and you want it in your name, well, you take the pine sale." So away they went and they took the sale and Dave cut the pine for the other guys.
>
> I put the pine on the inventory. And he had built up the inventory for all these other guys and they hadn't cut a stick. I said, that's a crooked deal, and this is the second time I goosed this guy, on quota. So I went up and called a meeting in Williams Lake and said I was going to put this volume of pine on the inventory. I was going to divide it up and give quota. But before I gave any quota out I was giving a quota to Dave Ainsworth based upon what he had. Before I started dividing everything I was giving him a quota off the top. If they had any complaints, let me hear them now. From that day to this I never heard a complaint. The whole industry said, Okay, give it, he earned it, that was fair. In those days in the Interior we could talk that way with the guys and they all said yeah, that's fair, away you go.
> —Ray Williston

were established) organized into associations. The seemingly impossible task of getting dozens of fiercely independent, often cantankerous and usually unco-operative loggers and mill owners to join an association and work together was accomplished by Larry DeGrace. He was a professional forester who started out with the Forest Service and ran the forest experimental station at Aleza Lake east of Prince George. He set up a forest consulting business in Prince George and single-handedly cajoled, persuaded, convinced and browbeat the quota holders to join and work through the operator associations. It was these groups which in practice determined the quota positions of the operators in each area, and maintained the stability of the system. They also made it harder for outsiders and nonmembers to acquire timber through the bidding process.

DeGrace also played a pivotal role in shaping the direction the Interior industry took after 1960. The bush mill and planer system was a prodigious and wasteful consumer of timber. To begin with, a substantial portion of every stand was left behind in the form of tops or trees that were too small for a bush mill to cut. At the mill, the sawyer began by squaring the log into a cant, slicing off four slabs which went into the waste pile. The circular saw blade usually cut a $3/8$-inch (9 mm) kerf, converting a large portion of the log into sawdust. Because the mills cut inaccurately, large tolerances were left to be taken off by the planer, producing more waste in the form of shavings. Old-time sawmillers had a joke about their business, part of which defined a sawmill as a device to produce sawdust, slabs and, as an incidental by-product, a small amount of lumber.

In 1956 Tom Wright, chief forester for Canadian Forest Products and a former dean of forestry at the University of BC, analyzed the output of the northern Interior forest industry and concluded that only 25 percent of every timber stand logged ended up as lumber. The remainder was wasted or, at best, used to fire the boilers in the mills.

A jammer used for loading trucks at Holding Lumber, Adams Lake, 1950s.
BCARS 73939

Pretty near everything that we tried out to see whether it would work or not, took place in Prince George because it was my constituency. I had to answer for it and I had to know individuals right there. And we had some wonderful people in Prince George to make this system start to work. One of the chief ones was Larry DeGrace. Larry had run an experimental station at Aleza Lake and then he came in and set up his consultancy, Industrial Forest Service, in Prince George. That man had no feet of clay. We were also setting, right at that time, Public Working Circles. Once we set up a Public Working Circle they set up an Association to represent the licensees within it. And every one of those associations in Prince George hired Larry DeGrace as their professional man. They'd always bring him when they came to me to argue a final thing in Victoria, and they'd be out there arguing a point and Larry would be there—and they paid him—and he'd say, "I don't agree with that. That isn't how it's gonna go." But they still hired him and they accepted him.

A loaded truck headed for a West Fraser mill, early 1960s. WFT

He was the only man in the province whose signature I'd accept as though it was signed by the Forest Service. His integrity was so absolute I never once had an objection from the industry, even though he might be right opposed to what they specifically wanted. His ethics in forestry were such that that's what he put down, and if he'd sign it, recommend it, I'd accept it.

When I put pine on the inventory it was about 68 or 70 miles from Williams Lake—a very remote area. Gabe and Conrad Pinette got together with this guy named [Rudy] Johnson and bought a bridge up in Alaska. They found a place on the Fraser River where they could stick this damn bridge across, and instead of being 68 miles away from Williams Lake, this timber sale would be about 16 or 18 miles away from Williams Lake, using the bridge. So they bid the timber in and then devised a way to dismantle the bridge up in Alaska and bring it down to Williams Lake, which they did. They painted all the side pieces and stuff and they put it together, and old Gabe strung cables across the river and they started building this bridge on one side. Hung it up on the cables and they kept pushing it out and then building it out there. I remember to this day getting a phone call in my office in Victoria and Conrad saying, "The bridge has landed." I often think of that. I knew all about this damn thing going across the river and I often thought, if one of those damn cables had snapped and that bridge had fallen into the middle of the Fraser River, what kind of hell would he have been into from that day to this.

They let all the farmers use the bridge for free. Anybody that trucked across it had to pay a toll to come across. They used that for many years and then eventually sold it to the Department of Highways. That's how Conrad and Gabe got a timber supply back again after they'd sold their other timber supply.

—Ray Williston

> By the time I was at Penny a lot of the river timber had gone. The best timber had gone and we were going up the valleys. We trucked down to the riverbank and decked our logs on the bank. One show I had at Penny was one of my brain waves. I got a timber sale up Driscoll Creek, which is right across the river from Penny Sawmills, and the grade was good. It was level, and then up Driscoll Creek. So I tail-skidded tree length with an old Dodge four-by-four. You put a bunk on the Dodge, chain the butts to the truck and away you go. We built an ice bridge and tail-skidded down, right across the Fraser to the mill yard. It was the cheapest goddamn logs I've ever delivered in my life, I think.
> —*Bill Batten*

The Prince George Board of Trade heard of Wright's study and hired DeGrace to investigate the economic potential of wasted wood. In 1960, for $200, DeGrace produced a report demonstrating that 35 percent of the waste produced in Prince George area sawmills and planers alone could be used to supply a 1,200-ton-a-day kraft pulp mill. As soon as he got a copy of the report, Williston stood up in the legislature in Victoria and, after quoting from it, predicted that pulp mills would be built throughout the Interior, supplied with waste wood from sawmills.

The owners and managers of the established coastal pulp companies scoffed. They used logs to feed their mills, running them through whole-log chippers. They burned their wood waste in their boilers, and the sawmills burned theirs in beehive burners, smothering Vancouver and the lower Fraser Valley in a blanket of smog. This is the way it is done, the coastal pulp people told Williston; it is ridiculous to think of building a pulp mill, let alone a whole pulp industry, on sawmill waste. A delegation of them visited him privately one night, trying to persuade him to drop the idea before he embarrassed himself. But as the next decade proved, they were wrong and Williston was right.

CHAPTER 9

Bigger Plans, Bigger Money

The 1960s

By 1960, enormous changes in the economic structure of the British Columbia Interior were underway. The booming postwar economy had gained a head of steam during the 1950s, and by the end of the decade the Social Credit government was heavily committed to a program of economic growth and industrial expansion—particularly in the northern Interior.

The province's financial resources, both private and public, were inadequate for the task of fulfilling Premier W.A.C. Bennett's northern vision. There were, however, large pools of investment capital available outside BC. This money proved critical to the spectacular growth the Interior forest industry experienced over the following decade. Big money was back, and although they did not know it yet, the bush mill operators were on the way out.

The first significant scheme involving large-scale foreign investment and BC resources was one advanced by Axel Wenner-Gren, a Swedish industrialist with vast financial reserves and an imagination to match. The BC government agreed to his proposal to develop 40,000 square miles (104,000 km²) in northeastern BC. It included a high-speed monorail line from central BC, up the Rocky Mountain trench to the Yukon border and, eventually, to Alaska. Other northern industrial projects, including a hydroelectric dam on the Peace River to power the railway, were also projected. Finally, an intention to develop the area's mineral and forest resources was expressed.

The Wenner-Gren plan, made public in the legislature by Bennett in 1957, was not merely another mad scheme to bilk BC of public funds. There was substance to the proposals, as there was to the government's response. The hydroelectric component, for instance, led to the creation of BC Hydro in 1961 and construction of the Peace dam, which was completed in 1967.

In a related development in 1961, Wenner-Gren set up Alexandra Forest Industries (AFI) to build an integrated forest products complex on the 260-mile-long (416 km) reservoir which would be created behind a dam on the Peace. Lacking the technical expertise and local connections to launch such an operation, Wenner-Gren engaged a coastal company, BC Forest Products (BCFP), to prepare a feasibility study. BCFP, in turn, brought in Larry DeGrace's Industrial Forestry Service in Prince George.

> Bill McMahn was retiring and I wasn't ready to take on the job that Bill had done. John Liersch agreed to come here from 1961 to 1971 at which time he would resign. He came on a ten-year contract. By that time we already had developed a fair industry in Grande Prairie, Alberta, and we had been negotiating a Forest Management Agreement to support the construction of a mill at Grand Prairie. We were satisfied that the timber was available but we had Howe Sound Pulp, which was a relatively modest-sized coastal mill, and most of our customers were offshore, so all our good contacts on pulp sales were in the overseas market. To build a pulp mill in Grande Prairie you had to be shipping significant quantities to the US market, where we were not well connected. So I went to see the government and told them that we were looking for a US partner but it would take some time, and without a US partner we couldn't undertake to do that because we were putting our integrity on the line and it would take us a while.
>
> This was at the end of 1960. On January 29, 1961 I was asked to speak to the Prince George Economic Development Commission, headed by Harold Moffat. On my way back from one of my frequent trips to Grande Prairie, I stopped in Prince George. The subject was pulp chips. Nobody was making chips in the Interior. Crown Zellerbach had taken the initiative, much to their credit, to offer to buy pulp chips here when the PGE went in, to bring them to the coast. They had offered a number of mills $5 a bone dry unit, FOB seller's mill. Well, the problem was that there were only two big sawmills in the Prince George area, Eagle Lake and Shelley, both of which at that time were owned by Binnie Milnar, who also owned the local newspaper and the local dairy. He correctly surmised there was no money in it to install a barker and a chipping and a conveying system in those mills because the $5 wouldn't give him a return.
>
> I was aware of that and I told them that the only way anybody was going to build a pulp mill in Prince George, after all these years of having wanted one, was if we could get the residuals. But residuals couldn't be produced from the bush mills and except for those two mills—one was 30 million a year, one was 60 million—all the rest of the mills were out in the bush and delivered very rough-sawn lumber to what they called Planing Mill Row. I suggested that if they had a local pulp mill they could expect about double the price and if they were larger mills, at that price, it would not only pay but it would make the mill better and give more year-round employment than what they had because the bush mills couldn't operate when it got really cold. It was a big front page story: "Coastal expert says ten dollars possible but it must be a local pulp mill." Then I got our forestry people to look at it, to the point that when John joined us in September of that year, he would be in charge of the project with his experience as a forester.
> —*Peter Bentley, chairman of Canadian Forest Products (Canfor).*

In the meantime other big-money interests were attracted by the increased activity in the north, and began to position themselves for construction of pulp mills. Canadian Forest Products (Canfor) was the first in line, acting on Tom Wright's report and DeGrace's wastewood study. It was a somewhat tricky time for Canfor. The company's co-founder, Poldi Bentley, was edging his way out the door into retirement and his son Peter was moving up to take the helm. But Peter had no experience with a development of this scale, so he brought in John Liersch, a veteran of the coastal pulp business. Liersch had been at the Powell River Company and was happy to escape the bloodbath about to take place following the merger-cum-takeover with MacMillan Bloedel. He and Peter Bentley negotiated the establishment of Prince George's first pulp mill. This was not as simple a task as it appeared. A firmly entrenched sawmill industry had long been accustomed to running the show around Prince George and was well connected to Forest Minister Ray Williston, who was MLA for the city. The locals were not about to be impressed by two city slickers from Vancouver.

It was not a simple task to shoehorn a pulp operation into the established Prince George forest industry. Hundreds of bush mills were still cutting rough lumber in the surrounding forests and delivering it to Planer Row in Prince George. And there were still three or four large mills operating on the East Line. Among them, these operations laid claim to the available sawlog supply around Prince George. The stands contained plenty of timber smaller than the minimum sawlog sizes, but no licensing procedure for making it available to a pulp mill.

Page 170: Schneider Logging's new heel boom loader at work near Kelowna, 1965. KLM

Page 171: A boomman at work on Okanagan Lake, 1965. KLM

Dumping logs into Okanagan Lake with an A-frame, 1965.
KLM 12,108

Consequently, Canfor concentrated on coming up with a mill proposal that would rely largely on sawmill residues. This involved working out firm agreements with existing Prince George mills, which was not an easy task and tended to make Canfor's lenders nervous. They didn't like the idea of the pulp company, a newcomer to the area, being dependent on a well-established, freewheeling sawmill industry that at some future date could hold the company hostage over prices.

At about this point in the proceedings, in 1961, a second big-time newcomer to Prince George began lining up a pulp mill. Noranda was a Toronto-based mining company wanting to diversify. It decided to get involved in the forest industry and, for openers, bought National Forest Products. National was a smoke-and-mirrors type of operation, put together by a promoter who bought several failing mills east of Prince George and in the Okanagan, milked

> I was Canfor's chief forester in 1961 when the company became interested in the possibility of building a pulp mill at Prince George. They sent me up to look over the timber and the industrial community there, the sawmills. And to see the possibilities of supplying logs and sawmill residues to a pulp mill. I spent quite a bit of time in the Prince George region observing the logging operations and looking at the timber and the timber types, the great spruce and lodgepole pine stands which surrounded the area. It was a colourful industry.
>
> The waste of the forest was enormous because of the lack of any market for the waste from the sawmills. The mills were generally built in the timber, and of course each mill had its own burner and the slabs and mill waste were sent up to the burner. The rough lumber was then shipped to the railway. In Prince George in particular there was an avenue called Planer Mill Row. There was a whole row of several planer mills. There were no logs delivered to Prince George, only rough lumber. Obviously you can get far more wood on a truck if all the slabs and trim are taken off. In the time that I worked in Prince George, I cannot remember seeing a single truckload of logs going into this enormous lumber manufacturing centre.
>
> —Tom Wright, *former dean of forestry at the University of BC and chief forester at Canfor. He and his son currently operate Witherby Tree Farm, a privately owned forest and Woodlot Licence on Howe Sound.*
>
> While we were preparing our proposal, along came Noranda, and Adam Zimmerman's first appearance on the scene. They bought a company called National Forest Products from Merv Davison. Merv was a promoter, he wasn't in the lumber business. And while they were going broke, he was building landing strips at his mill so he could fly his plane in. All of a sudden, instead of there being one remote chance of building a pulp mill in Prince George, two groups had expressed interest. Their group had already bought National Forest and we had done all the work with the Forest Service and with Ray Williston saying, we like the concept of working off residuals. I'm not sure they didn't also get one of Binnie Milnar's mills. Now they have both of the original mills.
>
> We felt that a backup guarantee was all that we needed. To make sure, because all the sawmill industry literally had to change from bush mill to centralized mills, one of the first people we met with was Nick Van Drimmelen of Netherlands Overseas. He had mills at Likely and other places, and he centralized in Prince George where their operation, which we now own, is located. Reed Paper and ourselves wound up as 10-percent owners each because of the financing we did to enable him to centralize all his production and start producing chips.
>
> One of the very first people to support us was Ivor Killy and his company. A number of people started to come on board that were the local lumber barons. They felt comfortable because the terms of a pulpwood harvesting agreement were not to conflict with but to complement the sawmill industry—and we were not to go into the sawmill business. When we went through the public hearings, we were challenged by Noranda, who said that we didn't have the wherewithal to do this. My dad spontaneously said, "We'll put up a performance bond to the government to undertake to build this, and if we don't build a plant of at least 500 tons by such-and-such a date, we will forfeit the money." That gave Ray Williston a chance to go ahead because he said the public interest was being protected.
>
> —Peter Bentley

> ▶ *A gathering of the major figures in the creation of Canadian Forest Products' Prince George pulp mill, 1960. Left to right: John Stokes (BC's chief forester), Larry DeGrace (Industrial Forestry Service), Ray Williston (minister of forests), unidentified, Poldi Bentley (Canfor), John Liersch (Canfor), Ralph Robbins (MoF), Fin McKinnon (MoF), unidentified.*
> Tom Wright collection

Pages 176–77: Schneider's A-frame dump on Okanagan lake, 1965. KLM 4296

as much cash as possible from them, and let them deteriorate even further. Noranda picked up the business and sent Adam Zimmerman out from the east to put it back on its feet—and get a pulp mill underway.

Noranda was in a different position than Canfor. It had timber tenures of its own, and captive sawmills to supply it with residues. But these supplies were not enough to run a pulp mill: Noranda too needed a means of ensuring a steady supply of wood. Zimmerman, who came from the heart of the Eastern establishment, was viewed even more suspiciously by the locals than Bentley, so Noranda hired Ian Mahood, a coastal forest consultant, to study its wood supply prospects. He came up with a brilliantly simple solution that enabled Williston to superimpose a new pulp industry on the existing sawmill industry, to the benefit of both.

> The lowest chip price we ever actually paid was $10.25. That was the initial price and it went up quickly after that. But at $10.25, and with the financing packages we offered people to consolidate, that was a better return on the chips for the added investment than the lumber was making on the sawmill side. Once you had a central sawmill and you installed a barker and chipper, $10.25 looked pretty good. And then it escalated and got tied to the pulp price and everything else and it got so high. Our theory was if the chip price plus the back haul of pulp from Prince George to loading it in a deep-sea ship in North Vancouver and Squamish offset the difference between the coastal chip price and the Interior chip price, we could be competitive with a coastal mill but we'd have superior quality. And that was the premise of building in the Interior.
> —Peter Bentley

Unloading chip trucks at Canadian Forest Products' Intercontinental Pulp mill, Prince George, 1984. CFP

Mahood's solution was called a Pulpwood Harvesting Agreement (PHA). It guaranteed pulp mills a supply of wood by permitting them to log timber below sawlog specifications, if—and only if—sawmills in the area were unable to supply sufficient residues. Stumpage on pulp wood was reduced, and pulp companies were not allowed to bid against sawmill companies at timber sales in the area. Any sawlogs harvested under a PHA had to be sold to the sawmills. On the other hand, in most of the agreements, sawmills obtaining timber from within the area covered by the PHA had to sell their residues to the pulp company holding the agreement.

Williston's intent was to set up a balance of economic power between the established sawmills and the new pulp mills, ensuring the latter got the wood they needed and the former got a fair price for their residues. Over the following decades both sides claimed they were getting the short end of the deal, but the system worked fairly well as a means of launching an Interior pulp industry.

> I went to see Martin Caine and ask for his support because he was kind of an elder statesman. He had Caine Lumber on Planer Mill Row. Mr. Caine said, Fella, where are you gonna build this thing? I said, Sir, right across the river from you right down there. I can show you out your window. Oh ho, he says, you guys are smarter than all these local bastards. He said, I've always told people there's gold under that property. So you're gonna get the property rights and then you're gonna mine it. You're not gonna build a pulp mill.
>
> He was an amateur prospector. He had rocks all over his office. I said, No sir, we're gonna build a pulp mill on it and we don't get the mineral rights. We're negotiating to get the property because we're reluctant to put that heavy an investment in leased land. So we're negotiating that with the province but we will not have any mineral rights on it, so I assure you we're going to build a pulp mill. Well, he says, maybe you guys are okay. If you don't get the mineral rights I can probably support that.
> —Peter Bentley

Canfor applied for and received the first PHA after Williston personally chaired a public hearing on its proposal at Prince George in 1962. The local sawmill community showed up in force to support Canfor. Zimmerman, representing Noranda, appeared and unveiled his proposal for a second Prince George pulp mill. He objected to the Canfor PHA application because it included areas Noranda wanted. A year later Canfor got its PHA and, in partnership with Reed Corporation from Britain, built the Prince George Pulp Company mill, which opened in 1966. In 1964, in partnership with Mead, a US paper company, Noranda got another PHA—the third issued—and built a mill under the name Northwood. In 1965 Canfor's Prince George Pulp, in partnership with the German firm Feldemuhle, got another PHA in the Prince George region and built the city's third pulp mill, Intercontinental Pulp.

> This was the political balance that Ray Williston had to deal with: on one hand he had independent sawmillers who came to him and said, we can supply the chips. Then he had the pulp mills saying, we've got to spend hundreds and hundreds of millions of dollars putting in these plants, we've got to be able to go to our bankers and say we have a guaranteed supply. We can't be held hostage. That was the other side of the argument. And I think that Williston, when he came up with the pulp harvesting areas, solved the whole problem. He was an astute enough politician to recognize the importance of giving the people who wanted to spend all of those huge amounts of money a guaranteed supply. I'm sure in the back of his mind Williston was smart enough to know that the whole thing would come out in the wash and there would be enough waste wood chips to satisfy everybody. History has proven him quite correct. I think the people in British Columbia owe that man a huge debt of gratitude. He never did get the proper recognition for that, I don't believe.
> —Fred Stinson

The floodgates were open, and new developments were hatched throughout the Interior. The introduction of new, big money into the industry, along with the growing market for sawmill residues, exerted great pressures for change on the sawmill sector. The industry was in a state of flux, but the policy structure within which it operated was sadly out of date. Current policies had evolved to accommodate a certain kind of sawmill industry, which was now undergoing a transformation; the policies were poorly suited to modernization of the sawmill industry and to the addition of a pulp and paper sector. The greatest difficulties stemmed from the fact almost all the forest land was owned by the government and the industry was dependent on timber from this land. The government was constrained to act within a policy framework which applied across the province, according to laws passed in the legislature. It was an inflexible system, particularly in regard to the various forms of tenure by which forest companies acquired government-owned timber. In the Interior, by 1960, the primary means of

Stacked pulp at the Intercontinental Pulp mill, Prince George. CFP

obtaining timber was still through short-term timber sales. Several Forest Management Licences had been awarded, but by 1960 they had been discredited due to the Sommers scandal, and few were issued after Williston became minister. He felt they were poorly suited to conditions facing the Interior industry in 1960. One of his solutions was the Pulpwood Harvesting Agreement, which solved the problem of superimposing a pulp sector on a sawmill-dominated industry. But other parts of the Interior had different problems that existing tenures could not solve. For instance, what kind of tenure was needed to induce the industry to utilize small trees? And what was required to encourage development in that half of the province lying north of Prince George, where there was practically no industry presence?

> Scaling wood had to change from board measure to cubic scale and then to weight scale. We made the three jumps there because when you started in on this small wood there was no way you could scale the damn stuff fairly. Then we went to weight scaling and how to integrate weight with volume—that took some more. I woke up to that with Crown Zellerbach down in the States, not up here. Before that they hadn't wakened up. Nobody paid any attention to the Interior so the Forest Service allowed the Interior to use the same scaling standards they used on the Coast. You only had to measure one end of a log, because they had a standard taper on the Coast. If it was a log so long and it was so big at the top end, it was so big at the bottom end. Soon after I got in there, we began checking up on shipments of lumber compared to the volume of stumpage they paid. Particularly a sawmill in my own district. They were shipping about one and a half times as much lumber as they were scaling in wood. We went to them, and no, they were scaling absolutely honestly. There was nothing to see. But you can't be, we said. So we took the suckers to court on this and they won because we hadn't wakened up. They were using the Coast taper, only measuring the top end of the logs and the taper on Interior wood is so much different than the taper on Coast wood. If you measure the top end of Interior logs, you're getting about twice as much wood. And on that he won the case! We didn't. So we had to change the whole scaling system.
> —Ray Williston

The question of the growth of the northern industry was dealt with during the evolution of the Wenner-Gren development in the northeast. After BCFP completed the forest inventory in 1964, it bought control of AFI from Wenner-Gren and applied for a Tree Farm

A heel boom line loader at work near Merritt, 1960s.
NVMA PL204

Licence. At about this point, several competitors appeared. Ben Ginter, a one-time Prince George logger and small mill owner who had branched out into highway construction, put in a proposal. Another offer came from Bob Cattermole, a Fraser Valley coastal logger. Cattermole had started logging in the north in the 1950s, bringing a crew and equipment north for the winter, and returning to the Coast to work for the summer. In the early 1960s he bought a sawmill at Fort Nelson. When the Peace reservoir began to be logged in 1963, Cattermole obtained several salvage sales and set up a series of bush mills at Finlay Forks, where the Finlay and Parsnip rivers join to form the Peace River. In 1965 he opened a temporary planer mill at Mackenzie. While working on the salvage project he realized, because of his coastal experience, that the true opportunity in this area lay with the timber that could be harvested along the huge, navigable reservoir once it was filled, and moved cheaply to a mill at Mackenzie. So he put in a proposal to build a pulp mill.

Williston resolved the issue by granting Alexandra Forest Industries 60 percent of the timber in the area under another new form of tenure, a Timber Sale Harvesting Licence. The remainder was sold at an auction in which Cattermole outbid Ginter. In partnership with two Japanese paper companies, he set up Finlay Forest Industries (FFI), which eventually consisted of two sawmills, a pulp mill and a newsprint mill. Between them, AFI and FFI built the town of Mackenzie, which was incorporated in 1966 and, in time, grew to a population of 5,000.

At the same time a northern pulp industry was getting off the ground, pulp mill proposals began to sprout in the southern Interior. Celgar, after consolidating its sawmilling operations at a new mill at Castlegar, opened a pulp mill at the same site in 1960—making it the first operating pulp mill in the Interior.

> A lot of people say the Williston reservoir was never logged, but it was. The specs at the time were a twelve-inch stump and a six-inch top, so some people think we didn't take anything at all.
>
> It was interesting. We felled a lot of timber, limbed it, and then as the water came up we gathered it together with boats. We had thirty-three boom boats working. That's how we would gather it together.
>
> We were able to export a lot of it. We sold lots of it into Grande Prairie to the plywood mills, and we had a whole bunch of bush mills of our own going, and one central planing mill.
>
> There was nothing here when we came. As soon as the power arrived we moved a camp in, then we moved our planing mill. We bought a little mill down at Kennedy siding on the railroad. Then when BC Rail came here in 1965 we brought the mill up by rail. By then we knew where the town was going to be, and we could choose the mill sites.
>
> —John Dahl, one of the founding managers of Finlay Forest Industries and a mayor of Mackenzie.

In Kamloops, longtime sawmiller Ken Long began working on a pulp mill proposal in the late 1950s. His first step was to organize several Thompson valley sawmills to ship residues to a coastal pulp mill. He almost succeeded in talking Crown Zellerbach, a coastal pulp company, into taking a piece of the action, but the company's head office in the USA would not permit it to get involved. Continued interest on the part of Crown's Canadian managers persuaded Long, Art Holding and Phos Bessette, owner of BC Interior Sawmills at Kamloops, to form Kamloops Pulp & Paper (KPP) and to assure Williston they had financial backing. He was convinced and in 1964 he granted them the second Pulpwood Harvesting Agreement issued. It included a deadline for startup of the mill, and when KPP realized Crown was not going to come in on the deal, they began a desperate search for another partner. They soon hooked up with Weyerhaeuser, one of the biggest US forest companies, which acquired 51 percent of the company and opened a pulp mill in 1965 with Long as president. This mill operated on sawmill residues from the outset, and Weyerhaeuser subsequently acquired a number of sawmills in the area, before opening additional Canadian operations in Alberta and Saskatchewan.

By 1958 Crestbrook Timber was verging on bankruptcy, after only two years of sawmill operations since its formation. Alf Farstad, the company's founder, and Vic Brown, an accountant who had begun with BC Spruce Mills at Lumberton in the early 1930s, managed to hold it together over the winter, and they enjoyed a spectacular surge in the US lumber market in 1959. Thus encouraged, they decided what they really needed was a pulp mill and they set out to find a partner to put up the money. After trying MacMillan Bloedel and a couple of other major companies without success, they heard that a delegation of Japanese business people was planning to visit the recently opened Celgar mill. With nothing to lose, they invited the delegation to Cranbrook and outlined their plans. The Japanese expressed polite interest and left. After a couple of years, having heard nothing, Brown and Farstad went to Japan and managed to fire up some interest. In the end, they persuaded Mitsubishi and Honshu to finance the pulp mill for only 48 percent of Crestbrook.

Farstad and Brown immediately applied for a PHA, which stirred up interest among other Columbia Valley sawmill companies. Williston decided to hold an auction for the PHA and the bidding narrowed to Farstad and Brown on one side and Kicking Horse Forest Products on the other. This company was a wholly owned subsidiary of Cypress Mines, a California corporation which had bought the Sigalet family's Golden Lumber in 1960. Williston got word that Kicking Horse was bidding only in an attempt to keep Japanese interests out of the Kootenays and had no intention of building a mill. When bidding reached an unrealistic level, he stopped the auction. Crestbrook went ahead with its pulp mill without a PHA, obtaining its wood supply in the form of residues from Kootenay area sawmills.

> Alf Farstad and I, we were real good friends and he was a fairly well-to-do guy, not a rich man. We had thought in '59, or even before that, that we should have a pulp mill instead of having all these damn burners around, to take all the waste. We started contacting people who might be interested in joining us. We didn't want to sell our outfit, we wanted somebody to build a pulp mill and we'd maybe own 52 percent of what we had. So we talked to quite a few people. MacBlo, they weren't interested in anything over here. They were quite interested in having a sales contract to sell our pulp. They weren't interested in coming into it. The American outfits, one or two I talked to, they'd buy the whole thing. That wasn't what we wanted.
>
> Then a friend of Farstad's phoned him one day and said there was a group of Japanese who were coming to Castlegar, which had just opened that year. He says maybe he could get them to stop over and talk to us if we were interested. Well, we were sure as hell interested. We had them to a dinner here and told them what we had and that we were interested in somebody putting up some money for a pulp mill and be maybe a 48-percent partner in the whole outfit. They did quite a lot of yakking and they were quite interested but they have to study things, you know. They said they would take it up with their company over there and see if they were interested or not. We never heard and we never heard and I was getting pretty fussed about it all.
>
> In the summer of the next year I said to Farstad I thought we should go and talk to those Japanese. "Oh, you can't do that!" He was quite a gambler. He said, "You'll weaken our bargaining position." And I said, We're not bargaining or doing a damn thing right now. Oh, well, he didn't think so. I said, I just think we should go over there. Anyway he finally said, Okay, okay. We'd better take our wives.
>
> My wife and I and Alf and his wife, we went over in August '64, late August. They were very interested. Alf and I went back again. They wanted us to present them with some kind of a proposal, so we did. Took it over and we went back again. I don't know the sequence but anyway we had a lot of meetings and finally we got to a go-ahead.
>
> —*Vic Brown. He was born in Ontario and started working for BC Spruce Mills in the mid-1930s. He retired as president and general manager of Crestbrook Forest Industries in 1979.*

The antagonism of US parent companies toward Japanese investment in the BC forest industry also surfaced during development of a pulp mill at Quesnel. Weldwood obtained a PHA in the Cariboo in 1965 and was in the process of forming a partnership with Daishowa, Japan's biggest pulp company, when Champion Papers in the USA bought Weldwood. Champion terminated the deal with Daishowa and put the mill on hold until Williston intervened and got it back on track.

Although most of the early pulp ventures came to fruition, there was one notable failure. In 1963 a group of Houston sawmillers formed Bulkley Valley Pulp & Timber and obtained PHA #4 on almost 10,000 square miles (26,000 km^2) of land at the headwaters of the Skeena River. They then cut a deal with Consolidated-Bathurst of Montreal and Bowater Paper of London to finance a pulp mill. These two firms bought control of BVP&T and, as a condition of government approval, agreed to build a pulp mill by 1973. A quick study showed them that a fully integrated complex was needed if the operation was to work at all. Since the PHA included no sawlogs, BVP&T began to buy local mills and their timber quotas. Among their purchases were Harry Hagman's Buck River Lumber and several mills around Babine and Burns Lake owned by Cooper-Widman, a Vancouver lumber-marketing firm.

In the late 1960s it was decided that sawmill operations at Houston would be consolidated and a pulp mill would be built there. A huge, ultramodern 200-million-foot-a-year sawmill was opened in 1970, and it was a disaster. The mill did not match the timber supply and the workers were not sufficiently trained to operate such a technically sophisticated operation. The pulp mill was put on hold. In 1972 Northwood bought the company and began shipping the sawmill residues to its Prince George pulp mill, thereby ending any chance of a pulp mill being built in Houston.

Big money arrived in the sawmill industry around the same time as it entered the Interior pulp sector, but to a lesser extent. Kicking Horse took over Golden Lumber in 1960, and

Noranda's purchases at approximately the same time, under the Northwood name, included several southern mills: Cooke Lumber at Greenwood, Western Pine Lumber at Princeton, Keremeos Lumber and Abernathy Sawmills at Keremeos, Hugh Leir's Penticton Sawmills, Osoyoos Sawmills, Summerland Box, Ellett Lumber at Beaverdell. Most of these mills disappeared as Northwood consolidated its southern operations. New mills were built at Princeton and Okanagan Falls in the 1960s, and both were later sold to Weyerhaeuser.

> Lignum had an operation in Salmon Arm, which was sold, that was a typical operation of the type where there was no quota attached. You had to be very inventive and very resourceful about how to get your logs and where to get your logs. Lignum applied in the Shuswap area for something called a cedar licence. For years and years and years, Federated Co-op was the main force in that area. And of course they were not interested in cedar or hemlock at all, they just wanted to have some nice spruce and fir. There was quite a bit of cedar and hemlock left over, which I thought at the time might be a good thing for the operation in Kamloops. So I put in an application on behalf of Lignum. At that time the regional forester at Kamloops was Al Dixon and he thought it was very amusing that somebody was coming to him for the cedar.
>
> He said, That's very interesting, I'll put it up. He did put it up, and I remember the auction. It was bid up to the point where it was not economical. Dixon, who was doing the auctioning, kept saying, What do I get now for this rotten hemlock and cedar nobody wanted for years and years and years? And it went up and up and up and eventually an outfit called Drew Sawmills got the sale. Drew immediately turned it back to the government and asked for relief and the Minister cancelled the sale because it was quite obviously spite bidding.
>
> That was the start of getting some cedar and hemlock out of the area, which nobody wanted for years and years and years. When you speak of cedar you've got to look at the quality. There is cedar in the Cariboo area too, and it's just bloody awful. You can't do anything with it. The cedar in the Shuswap area was much better. But there again, you had to be very versatile. You had to have proper sorting facilities, you had to have proper debarking facilities, and you had to have equipment which is suitable to cut cedar. I remember getting cedar logs to the Salmon Arm operation. They went onto the conventional head rig which we had. When you put the dogs down to dog the log, it just shattered! Just blew up! Because it was only a shell, but it contained the nicest wood and the clearest. Some of the operations, because they didn't have money to buy that specialized equipment, they made shingles or shakes.
> —*Sidney Eger, who immigrated to Canada in 1949 and began working in a Vancouver box mill. He retired recently as chief financial officer of Lignum Ltd.*

Similarly, when the US-owned coastal company Crown Zellerbach moved into the Interior in 1965, it bought the S.M. Simpson company at Kelowna, which had already taken over a number of Okanagan mills with deep roots in the community. In 1969 Crown bought the Smith family's Armstrong Sawmills, which had mills at Enderby, Vernon, Falkland and Armstrong. Armstrong's Vernon mill started out as Vernon Box & Pine Lumber, built by Jim Strother in the 1930s. The next year Crown bought the 16-million-foot-a-year Ponderosa Pine Lumber mill from Wilf Hanbury at Monte Lake. In the early 1970s, all these mills were closed and their production consolidated in new sawmills at Armstrong and Kelowna.

Also during the early 1960s, two small Okanagan companies went into expansionist mode. Gerald Raboch, who had taken over his father's Trinity Valley-based mills, formed a partnership with William H. Steele Lumber of Vancouver, a sales company. As Riverside Forest Products, they began buying mills around Vernon and Lumby, starting with Monashee Lumber, Fisher Planing Mills, Long Lake Lumber, half of Salmon Arm Lumber, Calvert Planing Mills, Kingfisher Sawmills, Barnes Lumber, Crossing Lumber, Taylor Brothers, Don Agur's mill at Summerland, Balestra & Schnyder and R.V. Schmidt. Most of these were small owner-operated mills whose proprietors had lost the desire or the ability to persevere in the industry that was shaping up during the 1960s. Some of them were getting old and had no successors; others were faced with tax bills they could not pay. The Riverside move was one event in the early stages of a massive consolidation that took place throughout the Interior.

Even under these circumstances, however, there were still opportunities for enterprising individuals to start small and build up their operations. Bill Kordyban got set up in Prince George in the early 1950s with a planer and a growing fleet of bush mills, under the name Carrier Lumber. He was one of the first to obtain a salvage sale during the clearing of the Peace reservoir, and among the few to make money on that project. Kordyban is one of the most original people who ever worked in the northern Interior forest industry, a business sector noted for its colourful characters. He operates according to his own unique set of philosophical, political and technical opinions. His technological ideas in particular made him a genius at building large, portable, highly efficient sawmills that could salvage timber which would otherwise be left to rot, and do it economically. The operation at Finlay Forks, set up to mill timber salvaged when the Peace reservoir was flooded, was his first project.

> At that time there were big mats of timber floating all over [Williston] Lake, and there were lots of complications because different companies had felled the timber and put some investment into it. It would float up to the surface and drift off down the lake into another company's operating area. The Peace Salvage Association was formed to kind of keep peace between the companies and sort out the problems—divide up the areas, and so on. They had some pretty exciting times, almost came to blows a few times. They did, for about half of one year, establish a boom right clean across the lake up at Lafferty Creek, that separated one company's logs from another. But eventually the boom gave way and there was one heck of a mixup. Some companies' logs went all the way to the dam, and of course Hydro would be dewatering them and burning them up.
>
> Then of course the problem was to get the logs out of these mats and separated into bundles of sawlogs, because they were all mixed up with the junk and debris. All kinds of experiments were tried on that. It seemed like the most successful was a crane on a barge that pulled the logs out and put them into a cradle on another barge. That was really Bill Kordyban's invention. Everybody told him it couldn't be done, but he put the crane on the barge. I think the first crane he had on a barge, he had down on Peace Arm somewhere and it was working fairly well, except that they left it. The barge sprung a leak, and when they arrived the next day it was tipping over. The crane went off into about three hundred feet of water. It's probably still there.
>
> When Carrier Lumber decided to put a mill up at Finlay Forks everybody told Kordyban there was no water. But he got an old willow stick and wandered around there, and sure enough he found a place. They drilled down there, and they had to drill pretty deep, but he got enough water to run washing systems for two sawmills and all the water he needed for the whole camp. That was built just below the flooding line, and they operated those mills until they built the Mackenzie mill. Just at the very last they were building dykes to keep the water out of the burners and everything else.
>
> When he first started, Kordyban had to bring the logs up to those mills quite a ways from the water's edge—it was about a mile down across this completely cleared area to get to where they were taking the logs out of the water. It was too far for the forklift to bring the logs up so they used trucks. Anybody else would have just loaded logs on trucks, but not Kordyban. He got some old rock-loading dumptrucks, took the rigs off the back and built racks on them. He trucked the full trees up to the mill crossways. When the trucks got to the mill they would back up to the log deck and dump the logs onto the deck. One-hundred-foot-wide loads, coming up the slope. It worked like a damn. Everybody else would have hauled them lengthwise, but not him.
> —*Harry Gairns, the retired president of Industrial Forestry Service of Prince George.*

Clearing the timber off the 663 square miles (1,658 km^2) to be flooded was something of a nightmare for most of those involved. Time was short and the salvage operation was overseen by the Forest Service bureaucracy. Getting logs out of the area before it was flooded turned out to be uneconomical in many cases. Alexandra Forest Industries attempted to truck logs from Finlay Forks to Mackenzie to build an inventory for the opening of its first sawmill, but found the venture too expensive. In the end the company logged timber outside the reservoir, close to the mill, for its initial supply.

> For the first while when the water was coming up it was quite exciting. We had crews up there at that time and I think the fastest that the water came up in the area that we were working in was about thirty feet in one day. We rescued the fallers out of treetops and everything else, because they would be in falling trees and they'd get caught by the rising water and couldn't get back to where their camp or boat was.
> —Harry Gairns

The Ainsworth Lumber mill crew eating lunch in front of the loading gates, early 1960s. A truck was backed under the stack of lumber, knocking over the wooden "gate" and dropping the end of the stack onto rollers on the truck deck. This simple device saved an enormous amount of labour, as the crew no longer had to load each board by hand.
Ainsworth collection

There were huge volumes of timber too small for the planned Mackenzie mills, or for portable salvage mills, and various means were tried simply to clear these areas of their forest cover. Long, heavy cables were attached to two D9 Cats and were used to knock down huge swaths of timber. BC Hydro brought in a gigantic tree crusher, which knocked down trees with a ram, then crushed them and drove them into the ground with enormous rollers covered with steel teeth. The device spent most of its time buried in mud, or being towed to solid ground with a small fleet of D8 Cats.

In the end, much of the timber was felled and left to float when the water rose. This created huge floating rafts of usable logs mixed in with debris, and separating the logs from the debris turned out to be a difficult, expensive task. For years, BC Hydro hauled millions of tons of debris out of the water and burned it on the beaches.

In the midst of all this confusion and chaos, Kordyban came up with innovative solutions to problems not normally encountered by loggers and mill operators. In the process he made a lot of money, with which he later built a large sawmill in Prince George. In the 1980s, Carrier utilized the knowledge it had gained in the Peace reservoir to undertake a salvage operation in a large area of insect-killed pine in the west Chilcotin. Again, specially designed portable mills were moved into the forest and the lumber was cut on site. It was a highly successful venture that, unfortunately for Carrier, ended in a legal dispute with the Forest Service over who was responsible for reforestation costs.

> I immigrated from Holland in 1950, and in 1951 I moved to Smithers and started working for a bush mill. In those days there were forty or fifty sawmills around Smithers and Telkwa and they were moved to the timber. So that's where I started working. I worked for about nine years off and on, fairly regularly in the bush for that same person. I started out horse skidding. That mill was cutting about 20,000 board feet a day and there were three teams, each with two horses for skidding. We skidded to a skidway, a couple of trees lengthwise, and we rolled them on there. Then a Cat would come along with a sleigh or a sloop and we rolled on about twenty logs that the Cat pulled to the sawmill. So that way you didn't have to move the sawmill as often.
>
> And then, in the spring of 1960, I started for myself. First only with two people, logging in the morning, sawing in the afternoon. It was just a small mill. We could set it up in about two or three days. We used to move an average of three times a year. It didn't take long. The sawdust was just blown on a pile. It had a blower, a big circular head saw—about a 52-inch head saw. That's the only machine we had for sawing. We didn't have an edger.
>
> First we would take a slab off the log, turn the log over, take another slab off. Then throw it flat, take off another slab and then start taking the lumber off. With two guys, we probably cut 2,000 feet a day. That was about all we could do. The average output was about a thousand board feet per man, I guess.
> —John Veerbeek, a former sawmill owner who logs in the Bulkley Valley.

By the early 1960s, Interior loggers desperately needed new equipment to deal with changing conditions in the woods. Workers were harvesting trees that were increasingly smaller, and farther from good roads, rivers or railway lines. Although chain saws had completely replaced hand saws and axes, more stems were needed to produce the same volume of timber, so a lot of fallers with a lot of saws were required to fall, delimb and buck the trees.

In 1960 hundreds of horses were still used throughout the Interior to skid logs out of the bush. By this time most skidding over longer distances—up to a mile or so—was done with crawler tractors, and arch trucks were generally used for even longer runs. The first rubber-

Cliff Falloon's first skidder, a Blue Ox, 1954. These early rubber-tired skidders were built by the Four Wheel Drive Motor Company and used conventional, front-axle steering. WFT

A Blue Ox skidder with a turn of logs in the Nicola Valley, late 1950s.
NVMA PL236

tired skidders began to appear in the 1950s. They were designed to skid tree-length logs longer distances and return for more at higher speeds. They were particularly useful on the type of rocky ground which quickly wore out crawler tractors. Machines such as the Westfall Performer, the Blue Ox, the BC-made Loggermobile and the Ontario-built Timberskidder were essentially stripped-down, compact, powerful trucks with rear-mounted winches. They were intended to replace the Cat, and although a few of them were used in the BC Interior, they could not compete with arch trucks in that area.

> When they first came, they came out to the woods to demonstrate the power saw. Two of my brothers-in-law, they were pretty good fallers, they done a fair bit. And they said, Okay, let's have a race. They sharpened up their saw real good to see who could cut the log the fastest. I think they beat them with their crosscut. Beat the power saw! Not for long. They wouldn't try it today. I think they ended up buying a power saw not too long after that. Same with the skidders. When they first came out with the skidders they were like a modified farm tractor. I still remember the fella that was doing the falling for us then. He was a Dutchman too. We were standing and looking at this mud hole, and he said, "If that thing'll skid these logs through this mud hole, I'll buy one." And the guy got on there and pulled the logs through that mud hole and the guy bought one. It's almost a toy when you look at it today but that's the way you start.
> —*John Martens*

But by the mid-1960s, a new type of skidder had come on the market that was designed to replace horses. Its frame bent in the middle and it was equipped with a powerful winch. It was highly manoeuvrable and could be driven right to the stump and return to the landing with

A Caterpillar with hydraulic shears mounted on it, working near Fort St. John, 1973. This device, and skidders mounted with shears, were the first falling machines used in the Interior. CFP

We got into these Blue Ox skidders. They came with a little Hercules six-cylinder inline motor, with a governor on it. The transmission was very poor. From first to second gear was too long of a jump. You'd get her revved up and when you tried to shift, it'd drop the engine way down. They wouldn't torque at all. We overhauled one of them that wore out. I think it cost us pretty near $800, which at that time was an ungodly sum. You could buy a whole motor for that. When the other one come due for a motor change we went down to old Chuck Leith and said, what do you got for a good used Ford motor? By golly, he had a car there that was wiped right off but the motor was pretty good. So we put this Ford motor in it. Well, it had no governor. You could wind that old Ford up to 4,000 rpm. They didn't last near as long as the Hercules, but you wind that up to 4,000 and when you dropped it from first to second, you were still at about 2,200, which is their high torque. Holy smokers! It made all the difference. And those old worm-drive winches, they were for the birds. We put TD9 winches on, it made quite a machine out of it.

We couldn't use them at Sovereign Creek. We tried them but there was quite a bit downhill and that timber up there was just fantastic spruce. There was spruce there with twenty-inch butts and it would have 150 running feet in it. Beautiful stuff! Just fantastic stuff. Well, you'd get three or four of these behind that Blue Ox, you could pull 'em. But you'd get going down these hills and then you'd come to a curve, the trees went straight ahead and took the Blue Ox with it. Roll it two or three times. Nobody ever got hurt.

So we bought a couple of these old army six-by-sixes. Put a big arch on the back, about like a D6 arch. Boy oh boy! We climbed a little better than 1,000 feet in a little over half a mile, which is quite a grade. But them old six-by-sixes, they'd hock her and they could come down. They'd floorboard them things and come across the bridge over Sovereign Creek and into the mill yard at quite a speed. And they'd let the winch go while they were still going, and them logs'd spread out and cover the whole mill yard. Spread 'em right out so that the bucker could get at 'em without doing nothing more.

—*Cliff Falloon*

a turn of logs, weaving its way through stumps, standing timber and other obstacles. The first two brands to appear were the Tree Farmer and Timberjack skidders. Perhaps their most appealing feature was that at the end of the day, on weekends and during longer shutdowns, they could be turned off and no one had to hang around to feed and care for them. They cost less than $10,000 each and they could move more timber over the roughest ground in less time than anyone had ever dreamed. By the end of the 1960s, most of the horses were gone.

Skidders came in the beginning of the '60s. I bought a skidder in 1964, a Timberjack. One guy had a Clark, that must have been maybe '63. I was the second one. Now you had to get experience too, you had to change your system. From skidding with the tops you had to skid the butts. Instead of falling on the skidway, you had to do it the other way because you had to get the butts to get the weight on the skidder. Too many accidents were happening because they pulled with the tops. When they made sharp corners, with the weight behind them, they rolled quite a few. That Timberjack, in a way I didn't like it. The skidder itself was all right but the winch wasn't a line winch. You always had to come to a stop, take it out of gear, and then you could use the winch. When you want to winch a load in, you had to take it out of gear and then use the winch. You couldn't use both at the same time. It had a wide gauge, it was really stable. You could skid on steeper sidehills if you had to go that way. But it was progress. You could skid farther. The main thing, you could get in the trees very clean because you could lift up the butts and just the small end of the trees were dragging on the ground. And you could take bigger loads. Another thing, it was a lot faster going back. It didn't make a difference even if you had to skid a kilometre. It was really progress with a skidder.
—*Fritz Pfeiffer*

In 1966 is when I bought the first skidder that was in Slocan, a small John Deere skidder. They were narrow them days and easily upset. You had to be very careful. I didn't roll it except once—but that was enough, because it was a very steep hill and actually the way I come down was steeper than the way I went up. We were skidding tree lengths at that time and it was on a right of way, so there was no real road. A real sharp corner and there was a big stump on the outside of the road there, and I knew what length of trees I could skid to get around that corner so it didn't hang up. This one day I had one longer than I realized, I guess, and it caught on the stump and all of a sudden it let go and the damn machine went right over! Right over! The back wheels went right over the front wheels! End over end! How the hell I'm here today is more than I'll ever know.
—*Harry Meyers. He logged in the Slocan Valley from 1947 to 1970, then went to work for Crestbrook in the east Kootenays until his retirement in 1989.*

It was not much money, and for many years I didn't know if I was coming or going. It was tough. But I survived by hard work and perseverance, I guess, and there was one item that was our saviour. We cut scaffolding, two-by-nine, thirteen-footers and some ten. They went to the New York area. That brought quite a bit more money. If it hadn't been for that, we could have never survived. And that continued as long as I was sawmilling. I sold two-by-nine for a total of twenty-five years, from 1960 to 1985. I always cut two-by-nine for scaffolding and the only place it's pretty well used is around New York. It's used for construction, for scaffolding. It's an odd size and that's why they pay more. Anyway, that saved us.
—*John Veerbeek*

Hard work's never hurt nobody. Thinking about it has. You take falling. You still can't compete with the new machine with a hand faller for price—if that hand faller will work. But they got so independent and so snarky that everybody used machinery. Every time we've mechanized—I don't care if it's a falling machine, delimbing, skidding, the whole works—every time you get a piece of machinery, up goes the price. You can skid a lot cheaper with a horse than you can with a hundred-thousand-dollar skidder. Per metre. As long as you can find a guy to look after the horse. Knows how. And stay the weekends and feed him. Yeah, it's a very romantic thing until you find everybody else is having romance on the weekend and you're in camp feeding the horses.
—*Archie Strimbold*

A Drott feller-buncher working near Fort St. James, 1980. This machine revolutionized mechanized falling. CFP

Another machine which began to appear about this time was the first type of fully mechanized faller. This early, crude falling machine consisted of a large hydraulically activated shear mounted on the front of a Cat or rubber-tired skidder. Because the machine equipped with the shear had to be driven up to each tree, it was slow. With practice an operator could control the direction of fall, but this was not a very precise tool.

In 1968 a revolutionary new kind of mechanized falling machine was built by a Wisconsin builder of logging equipment, Erv Drott. He mounted a hydraulic shear and grapple on the end of a long knuckle boom that sat on an excavator base. This machine would sever a tree, control its fall and stack it in a pile. It was the first feller-buncher. Its most important capability, perhaps, was its ability not to fall trees, but to bunch them. Then, with the addition of a large grapple hung off the rear of a skidder, a dozen or more whole trees could be picked up by a single worker—who did not even have to get off his machine—and hauled out to the nearest landing. The combination of these two machines constituted the world's first fully mechanized logging system. It was the ideal combination of machines for the BC Interior, and they radically altered logging there over the next decade or two. Utilization of small logs and the development of equipment to harvest them led to widespread abandonment of selective logging systems, and the adoption of clearcut harvesting.

By the late 1970s, logging was as highly mechanized in the Interior as any place in the world. To a great extent this transformation was encouraged by an equally radical transformation of sawmills, much of it centred around a new type of saw developed in BC.

CHAPTER 10

Technological Transformation

The 1970s

The need for technological change in the Interior forest industry peaked in the mid-1960s, with the introduction of pulp mills into the economic equation. The mills' demand for fibre not only fostered a consolidation of the sawmill sector, but also led to the development of radically new methods of producing lumber from logs in order to provide pulp mills with more usable by-products.

The sawmill industry which existed in the late 1950s and early 1960s was poorly equipped to supply residues to pulp mills. The pulp sector required clean, bark-free wood, preferably in the form of chips. Larger, stationary mills could produce enough of these chips to make it economical to install debarking equipment and chippers—during 1965 alone, fifteen to twenty of the bigger Interior sawmills installed this equipment—but much of the sawmill sector consisted of small bush mills, hundreds of them, scattered far and wide. Their residues were in the form of sawdust, and slabs with the bark still on. Making these materials available to pulp mills was physically difficult and economically out of the question. Consequently, most of this wood was burned or left to rot in piles scattered throughout the woods. The solution was to consolidate the sawmills.

At the same time, a potentially larger source of supply existed in small trees which were not logged under prevalent utilization standards. Until 1966, Forest Service regulations stipulated that only those trees 11 inches (27.5 cm) in diameter or larger must be harvested and that tops could be cut off at a 6-inch (15 cm) diameter. In some Interior forests loggers were not permitted to cut trees smaller than 16 inches (40 cm) for silvicultural reasons, but otherwise these "intermediate utilization" standards, as they were called, acknowledged the inability of the mills to mill lumber economically from small logs. Most Interior stands contained considerable numbers of trees below this standard, and in the lodgepole pine forests most trees were below the standard. The solution to the problem of utilizing this wood lay with innovations in sawmill technology.

Beginning about 1965, consolidation and technological innovation transformed the Interior sawmill industry. Practically all the bush mills disappeared and were replaced

by fewer, but larger, stationary mills, and the established big mills were radically modernized. A measure of the effects of restructuring: between 1962 and 1971 the number of operating mills in the province declined from 1,627 to 627, while lumber production increased 49 percent.

To encourage consolidation, the government, in the person of Forest Minister Ray Williston, adopted several new regulations. For example, Williston permitted timber quotas to be sold, enabling mills to secure the increased timber supplies they needed to obtain financing and justify the building of bigger, more expensive mills. Two to three million dollars was being spent on sawmill construction projects in the mid-1960s, more than ten times the maximum amount spent a decade earlier.

After observing the success of the Cariboo mills' handling of small logs, Williston imposed

Page 192: The prototype of the Chip-N-Saw in Shelton, Washington. Its inventor, Ernie Runyan (left) and Canadian Car Company's chief engineer, Len Mitten (right), were the developers of the most revolutionary advance in sawmill technology in the twentieth century.
USNR

The Pas Lumber Company's Kerry Lake sawmill, 1963. Lumber from this mill was trucked to Prince George for planing. TPL

"close utilization" standards on the industry, requiring Interior loggers to cut to a 7.1-inch (17.8 cm) diameter and a 4-inch (10 cm) top. This measure alone increased the Interior's merchantable wood supply by 30 percent, and licensees who could demonstrate their ability to process the small logs had their quotas raised proportionately. Low stumpage rates of 55 cents a cunit (100 cubic feet/3 m³) on this small wood encouraged licensees to use it. After a few years it was apparent that even with these increases in the amount logged, additional harvesting could take place within the sustainable cut levels of most sustained yield units. Licensees with sawmills in areas where 80 percent of the logging was done to close utilization standards were awarded even more quota, known as third-band timber volumes. By 1975, 40 percent of the Interior timber harvest was third-band timber.

> After the war there was a wave of gang saws that came through the province. I think eventually we got up to about eight hundred gang saws. This is a sash gang. They had multiple saws and a reciprocating sash. The log goes through and it's cut into two-inch widths. By this time they were part of an element in a sawmill and they had a log haul, a millpond and so forth. So a gang saw followed by an edger and then a green chain. They'd pull pieces of various thicknesses off the green chain and resaw them. That went on for quite some time, but as the fibre became more and more valuable the problem with the gang saw appeared. You couldn't saw for grade. You couldn't take the clear lumber off the log because the log is tapered, and with a gang saw you just had to cut it into laminations, parallel laminations, and a lot of the clear lumber was going out in slabs. If you cut with a band saw head rig, then you can follow the clear lumber down the side of the log, then turn the log and cut the clear lumber off the other side.
> —Len Mitten, who graduated in mechanical engineering from the University of BC in 1946 and worked as chief engineer at CanCar during development of the Chip-N-Saw. He was a founder of Cetec, which developed the Smart Saw and the Rim Saw.

When large volumes of small logs began to come out of the woods and scaling the small logs became problematic, a weight scaling system was introduced. Instead of measuring each log, as had previously been done, truckloads of logs were weighed and timber volumes calculated. Since the mills were paying stumpage for all this wood, as well as paying to log it, they had a strong incentive to use it and recover their costs.

The machine that enabled the technical transformation of sawmilling was the Chip-N-Saw. It was developed at the Canadian Car Company's (CanCar) Vancouver factory in 1963, based on a crude prototype designed by Ernie Runyan, an employee at Simpson Timber in Shelton, Washington. The saw consisted of a single unit combining several chipping blades and circular saws. The first set of chippers cut a two-by-four-inch spline into the bottom of a debarked log, which served as a guide for subsequent chipping and sawing operations. Another set of chippers flattened the sides of the log, followed by a third set which chipped a spline on the top. The log then went through two sets of saws which cut it into 2-inch (5 cm) boards.

Page 193: An employee sharpens a saw blade at a Slocan Forest Products mill. SG

It was an impressively efficient saw, producing very little waste. Sawdust production fell from as much as 25 percent in conventional sawmills to about 7 percent. Some 60 to 70 percent of the log ended up as boards, and the remainder as chips. Only one worker was needed to operate the saw in a single continuous operation, as fast as logs could be fed into it. The first Chip-N-Saws processed logs at about 90 feet a minute, compared to about 20 feet a minute for the gang saws of that era.

The first Chip-N-Saw built was refined in a test lab at Canfor's Eburne sawmill in Vancouver. When Canfor declined to put it into production, it was sold to John Ernst, the outspoken Quesnel newspaper editor who owned Ernst Lumber. It was an immediate success, and CanCar began to turn out more, adding improvements. S.M. Simpson at Kelowna took

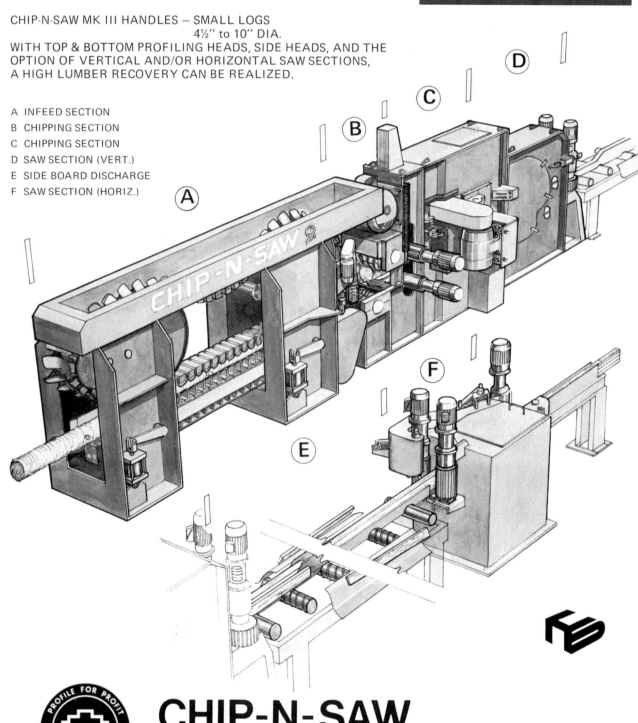

◀

*An early schematic drawing of a Chip-N-Saw. Peeled logs are fed in at **A** and pass through two sets of chipping knives in **B** and **C**. A vertical gang saw in **D** produces several boards, some of which are removed at **E**. The remaining boards are cut to dimension with a set of horizontal saws in **F**.* USNR

the next one, and then Sam Ketcham at West Fraser discovered the Chip-N-Saw. He ordered the next four or five units to come off the production line and couldn't get them fast enough. Canfor installed one in its Fort St. John lumber mill in Chetwynd, which it had just bought from Gordon Moore.

After that, Chip-N-Saws spread into practically every mill across the Interior, and throughout North America. Established mills used them in small-log lines they set up alongside their conventional mills. Beginning in the late 1960s, a lot of new mills were built around Chip-N-Saws. CanCar made hundreds of them in Vancouver before it finally shut down the factory and moved it to the USA. A couple of competitors came out with similar models, but backed off in the face of patent infringement suits.

> I first heard about the chipping saw when we worked with Eburne sawmill. I knew the people there quite well. They had a test lab, so if we were bringing out a new machine at CanCar, we would arrange to test it in their test lab. One of the electrical people there, Bill John, had been down to Shelton, Washington, and he met the electrical inspector down there. This fellow had started to put together a chipping saw in an old hangar, so Bill went and had a look at it and he thought, boy, that's what we need. Something to chew up those treetops that we're leaving in the woods. He got hold of me and said, Len, you gotta go down there, have a look at this thing. Well, I was chief engineer at CanCar Pacific at that time so I went down. I liked the idea just over the phone and I cut a deal with this inventor to take the rights for Canada and the USA on the basis that we would do a proper machine design. He couldn't keep this machine running for more than about a half an hour. It was sort of put together with bits and pieces. It had five thousand parts in it, so there was always one part that was failing.
>
> His name was Ernie Runyan. Just a natural inventor. A very logical mind. Decided this was crazy having all these acres of machine tools to break down a log into lumber. He worked for Simpsons, so of course he went to Simpsons and said, I've got this idea and I'd like to sell you the rights to it. They came and looked at it and of course it broke down during the demonstration. They all waved their hands at it and said, That thing'll never work. Forget it! You're wasting your money, Ernie. So I arrived at a propitious time. He gave us an option for six months, which I didn't even pay for. During that time we put together a test unit and, with Eburne's help in their test lab, tried it out there. The thing worked beautifully.
>
> Sam Ketcham was one of the first people to recognize the value of this, the economic value, so he bought a chipping saw, our third or fourth chipping saw I guess, and put it in. We used to say we'll make you a millionaire in five months if you put in a chipping saw. And that's about how good the economics were. He tried it out for five months, then he bought his neighbour's mill down the street. Williams Lake. And put a chipping saw in there. I think in all he bought four or five chipping saws from us. Sometime after that he was killed in a helicopter crash and I think his estate was a hundred million dollars or so. Getting there from a hand-to-mouth operation in just a few short years, he owed to the chipping saw.
> —Len Mitten

When Interior mills started moving into lodgepole pine in a big way, CanCar came out with a modified version of the Chip-N-Saw to cut lumber from logs as small as 4 inches (10 cm) across. Instead of cutting a guide spline, the new model had V-shaped chipping heads on the top and bottom which converted the log to a square or rectangular cant. The V-shaped bottom of the cant served to guide it through the remainder of the milling operation. This model operated at even higher speeds than the original Chip-N-Saw.

One of the major benefits of the Chip-N-Saw was that it greatly facilitated the development of continuous-flow sawmilling. Older-style carriage mills converted a log to lumber by running the carriage back and forth through the saw until the job was finished. Then another log was loaded onto the carriage and the process was repeated, creating an intermittent movement of logs into the mill. Sophisticated versions of carriage mills, employing double-cut band saws and other innovations, were efficient on large-diameter logs and well suited to the production of high-value clear lumber. But it takes almost as long to cut up a small log on a carriage mill as a large one, and when the time came to begin using smaller timber, carriage mills lost their place. Gang saws maintained a continuous flow of logs, but they were relatively slow.

With bigger, faster motors, Chip-N-Saws soon were processing logs at a rate of 300 feet a minute, creating a need for new high-speed milling equipment at other points along the production line. Higher-speed saws, debarkers and sorting machinery evolved quickly. Len Mitten, CanCar's chief engineer during development of the Chip-N-Saw, eventually went off on his own and started a company which produced two narrow-kerf, high-speed saws, including the Rim Saw and the Smart Saw. For a decade or more, largely in response to the needs of the Interior sawmilling industry, several BC manufacturers developed innovative milling equipment that found markets far beyond provincial borders.

> Doug Floyd was a tremendous operator. He and Sam Ketcham were the team, and they hired a guy named Ken Seaman. Hired him away from down in the Kootenays. Ken basically designed and built the first high-efficiency, small-log sawmill in the Interior. That was our mill in Quesnel that we ended up really going out on a limb for and used modern technology—chipping saws and canters and band saws. That was the mill that really gave us our head start on the rest of the industry because we were able to combine a couple of mills into one. It was highly efficient, state of the art. It was for a long time the most productive mill up in the Interior.
>
> We put a nine-inch canter chipping saw in, which was pretty small for those days, but it coincided with the small wood that became a requirement to harvest up there. They moved very rapidly after that. The first mill was a watershed because we didn't have very much money in the company. I think when we put in the mill in Quesnel they had to borrow a fair bit of money to do it, and I think they were worried about the startup. But once that thing went, it just never looked back. We made a lot of money off that mill and were able to duplicate it in basically every acquisition we made after that. First Williams Lake, then Smithers, then Fraser Lake and on and on. They were basically modelled after the mill we had built in Quesnel in '69.
> —Hank Ketcham, president of West Fraser Timber Company.

The new generation of sawmill technology, combined with increased revenues from chips flowing to new pulp mills, hastened further consolidation of the industry. During the late 1960s two more pulp mills were built, creating even more demand for chips. One of them was Cariboo Pulp at Quesnel. The other was built by Eurocan Pulp & Paper on the coast, at Kitimat, but got 50 percent of its timber supply from the Interior west of Burns Lake. It was financed entirely by Finnish interests, which took over the project when Alcan, the aluminum company that built Kitimat, and the Powell River Company, part of MacMillan Bloedel, backed out. The long haul from the woods to the Kitimat mill made this an expensive operation, as Ike Barber, the company's first vice president of logging, soon found out. Logs were trucked from the woods to a sorting ground, where they were bundled and loaded onto off-highway Hayes trucks and hauled to Ootsa Lake. They were towed more than 50 miles (80 km) to Tahtsa Lake, reloaded onto trucks and hauled another 35 miles (56 km) to tidewater. They were dumped, then towed another 50 miles to Kitimat, where they were dewatered and trucked to the mill. All the logging was done by Interior contractors on stump-to-dump contracts that were instrumental in introducing big-time, full-phase contract logging to the Interior.

In the sawmill sector, the fate of Boundary Sawmills typified the new forces affecting the industry. This mill was started at Midway in 1935 by John Sherbinin and his son, also named John. Their wood-fired, steam-powered mill could cut 35,000 feet a day and they struggled through the tail end of the Depression producing lumber for the local market and timbers for

Removing cables from log bundles which have been towed down Williston Lake to the BC Forest Products mill at Mackenzie, early 1970s. MDMS

We started work for Eurocan in '68 and worked there 'til '74, I guess. We built the roads and so on. And then we tried forwarding with a 980. Packed logs out of the bush. Then we bought skidders. So that's where we got quite serious into the logging instead of just something to do in the winter. Nobody knew what a skidder could do or what they couldn't do. We had skidders broke in half and upside down and inside out. I'll always remember one young fellow, and me giving him shit for having an arsehole in the main line. I put a new line on and what does he do? He does the same thing. He hooks on a stump, takes a run at it, gonna straighten this out before I catch him. Bust the hinge pin and the skidder's in two pieces! Another guy's stuck and hooked onto the canopy post. Pulled the radiator and everything out of the skidder. You know, like all of a sudden there's fifty jobs up here and twenty people to do them. Hire anybody that was warm. Yup.

—*Archie Strimbold, who grew up in the Bulkley Valley, started logging when he was fifteen and retired more than fifty years later.*

the CPR. Keeping their head above water at this time was difficult, and to keep the cookhouse operating the Sherbinins went into debt with the local grocery store in Midway. The store owner, Gordon McMynn, took his payment in the form of company shares and soon became actively involved in Boundary's day-to-day operations. The Sherbinin–McMynn partners built another mill in 1936.

After the war the company increased its output to 30 million feet a year by taking over several smaller local operations, including Grand Forks Sawmills, Fitz Sawmill and Olinger Sawmills. It obtained one of the first Tree Farm Licences and was renamed Boundary Forest Products in the mid-1960s. But in 1967 it went through a long strike and emerged half a million dollars in debt. By this time, in order to handle the small logs falling within close utilization and third-band timber standards, Boundary required a few million dollars' investment in new mill and logging equipment. This was a lot of debt for a family business to assume, so the owners elected to sell. Their company was purchased by Pope & Talbot of Portland, one of the oldest forest companies in the Pacific Northwest. New money was poured into the mill and George Sherbinin, the grandson of John Sr., stayed on as president for five

Holding Lumber's sawmill, Adams Lake, 1960s.
BCARS 73918

years. He then resigned to run S & O Sawmills at Westbridge and a wholesale lumber firm in Midway.

In a similar manner, Evans Forest Products moved into the Interior. Evans was an old Oregon company which began in the 1920s manufacturing batteries, then built a sawmill to cut yellow cedar for use in its batteries. In 1928 the company bought another mill in Vancouver to cut battery wood and lumber. In the late 1960s it launched an expansion into the Interior with the purchase of Savona Timber on Kamloops Lake, Commercial Lumber in Lillooet, the Kicking Horse mill in Golden and Selkirk Spruce Mills at Donald. Although its Interior operations did well, Evans's parent company in the USA went broke in 1986. A group of senior managers took over the BC division and consolidated its operations in a sawmill at Donald and a plywood plant at Golden. Ten years later, with a reduced annual timber harvest, the Donald mill was closed and Evans was in financial difficulty again. The partnership group was bought out by Lyle St. Laurent, a US banker. A laminated veneer board plant was added to the Golden plywood plant and, with financial assistance from the provincial government, the operation was put back on its feet.

Numerous other corporate transformations took place at about the same time. Holding Lumber at Adams Lake, whose founder Art Holding was one of the originators of Weyerhaeuser's Kamloops pulp mill, was sold to Whonnock Industries, a coastal lumber producer and a predecessor company of International Forest Products controlled by the Sauder family. Netherlands Overseas Mills had evolved from a mill built at Lone Butte by Nick Van Drimmelen in 1955. He took over several mills at widely scattered locations in the Interior, then consolidated his operations in Prince George in the 1960s and built a big, modern mill that went broke in 1967. A year later it was bought by Balfour-Guthrie, a Vancouver lumber wholesaler, and revived. In 1989 Canfor took over Balfour-Guthrie and incorporated the Netherlands mill into its operations. Wilder Brothers Lumber at Radium was bought by an Alberta company, Revelstoke Sawmills, which then bought several other mills in BC. Dozens of transactions of this type took place during the consolidation phase.

A few companies followed a different growth pattern. When a proposal for a big new sawmill at Burns Lake came up during the New Democratic Party's term of office, the government insisted on the participation of Native British Columbians. The result was Babine Forest Products, a highly successful joint venture between Weldwood, Eurocan and the Burns Lake Native Development Corporation.

> In 1977 the Tackama operation burned. Four months from the day that it burned we took the first veneer off of the first line that we had reconstructed. [Right at the end] we couldn't get the damn thing going. We were in Prince George with an old Cheyenne airplane and we got a phone call saying the card and clipper isn't working. So at 5:00 at night we're at the airport in Prince George, fly to Vancouver, pick a guy up, fly all the way back to Fort Nelson and we got there at 11:00 at night. I said, We're going to the plant and we're going to plug the card in because it's four months to the day that it burned. We can do it. And sure enough it worked—we got veneer off the line. Not very much, but we got it. To this day we can truthfully say we got veneer off the line four months from the day that the plant burned down.
> —Don Clutterham, operations vice president of the Slocan Group.

By 1965 the Lavington Planer Mill, built by Harold Thorlakson a few years earlier, had grown slowly and established a solid customer base. Thorlakson and his son Doug ran the operation with two portable sawmills, and his other two sons, John and Al, went to university to study engineering. Al spent a few years working at Quesnel for Western Plywood and its successor, Weldwood, then returned to the Okanagan in 1967. Having seen how lodgepole pine was being used in the Cariboo, and how mills in that area were making the transition to

The Thorlakson family in the early stages of Tolko Industries, 1960s. Al, the current president, is at left; his father Harold is at centre. TI

close utilization, Al realized that because most Okanagan mills were still working to intermediate utilization standards, there were good opportunities for growth. The Thorlaksons put a barker and chipper in their Lavington operation, then got a small third-band timber allocation. John joined the business in 1970 with a masters degree in engineering from McGill University. The family started cutting for value and in 1970 put in a dry kiln. Two years later they bought the Dumont family's sawmill at Newport Creek (known as the Hoover Sawmill, after its founders), and more than doubled their output. They changed the company name to Tolko Industries in 1972, when Harold turned the business over to his sons, and they began an expansion program which was still going strong twenty-five years later.

Centralization of sawmilling in larger, more efficient mills could not have taken place without concurrent changes in the logging industry, which had begun to emerge as a major factor in the Interior forest industry. In most parts of the Interior there had always been a few loggers who did not own mills and who undertook various phases of logging timber belonging to others, but this had not been the norm. During the bush mill era, most of the small mill owners did their own logging. When they sold their logging quotas and permanently parked their portable mills, many of them became full-time loggers. They invested the money they received for quota in new mechanized logging equipment and took contracts to log for the quota holders.

Cliff Falloon, who formed Quadra Logging at Quesnel, was typical of those who made this transition, as was Howard Lloyd, a Prince George mill owner, and several hundred others throughout the Interior. The Interior forest industry had not started out with separate logging and milling sectors; the division occurred at this time.

The specialization of the logging industry coincided with its accelerated mechanization. When the sawmills moved out of the woods to central locations, it was no longer possible to skid logs from stump to bush mill with a team of horses and haul the rough lumber to a planer with a two-ton truck. Now it was necessary to haul logs to the larger, centralized mills. It was at this point, in the 1960s, that conventional logging trucks capable of hauling tree-length logs

When I first went to Fort Nelson, the oil people would build ice bridges and they'd lay logs crossways to a roadway. They'd flood water on it with little pumps. It would take weeks and weeks and weeks to build an ice bridge and it would use two or three thousand trees to go across five or six hundred feet of river. There was a bunch of arguments between the government—Ministry of Environment, the Ministry of Forests—about putting all this wood in the river. In the spring it would just wash away. So one year we tried building bridges without wood in them. It worked but it was a slow, expensive process. At one point somebody that worked with us happened to see on a TV show that in Quebec, when they put wood out on the lakes in the springtime, they use auger pumps. You drill a hole in the ice and the pump is an engine like a Skidoo engine with an auger on it. It pumps thousands of gallons of water out through a chute onto the ice. We thought that was a heck of an idea, so we started using those and putting small logs lengthwise, parallel to the road, to act as rebar, spreading them apart. We found that was a pretty successful way to build ice bridges. In ten days and the proper weather, by using these high-volume pumps and continuous flooding and dumping snow, we could build thirty to thirty-six inches of ice on the river. It creates a floating pad on the water so that it will support the load but also floats it. There were very critical speeds to cross these bridges. They were limited on speeds because the weight of the load creates a deflection into the water and it puts a wave in the ice in front of the vehicle that's travelling on it. If you travel too fast, you climb over the wave and break the ice. Over the years a number of trucks have broken through the ice. In essence, an ice bridge is just a pad of ice built across the river. And with a meandering river, rather than try and build roads along sloughing cutbanks and so on, it was easier to build a bridge across the river and go across to the next point and build another bridge. Thirty bridges to go twenty miles, type of thing.
—Don Clutterham

Crossing Williston Lake on an ice bridge, headed for BC Forest Products' mill in Mackenzie, 1970s. MDMS

In '73 to about '79 I had about twenty-four men and I put out sixteen or eighteen loads a day. And then I had subcontractors too sometimes. Stump to dump. I used to do everything. Trucks. Built my own road, the whole works. Loading, grading, the whole thing. And once you done that, then the bigger outfits, they liked that 'cause then they could kind of forget about you. 'Course you didn't move that much then, either. When I logged for Kootenay Forest, you just stayed maybe four, five, six months in one place. Maybe move twice a year, that's all. Now they got postage stamp-sized settings. Trying to make it look good. You might as well forget about it. It was a rough old life, but it's a good life. The best years were the '60s and '70s. In the '80s they had that downturn, but then it got going again. It wasn't bad. You know, sometimes I wish I'd of kept a couple of pieces of machinery around. I could just get out there... ach, this retirement is all right but you gotta try to keep busy. When you're used to logging, you work at it and you get used to that and it's hard to get away from.
—Bill Seafoot, who was a contract logger throughout the Interior and, briefly, on the Coast, for more than forty years.

Following pages: Lifting log bundles from the dewatering elevators at BC Forest Products' Mackenzie mill with a LeTourneau log stacker. The elevators were used to raise logs from Williston Lake up a steep slope to the mill yard. MDMS

began to move logs from the landings to the mill. More and more trees were felled mechanically, either with shears or feller-bunchers. Rubber-tired skidders dragged bunches of whole trees to the road, where they were limbed and bucked with chain saws and, later, various types of delimbing machines and slashers. The old short-log loading systems did not work with longer logs, so new kinds of hydraulic heel boom loaders were employed to pile the logs on trucks, and huge stackers unloaded them at the mill yards. A decade earlier, the well-equipped Interior logger, with a power saw, a team of horses, a small truck and maybe a small crawler tractor, required only a few thousand dollars to set himself up. By the late 1960s, Interior contract loggers were investing a few hundred thousand dollars each to get into the business.

A 120-foot Douglas fir pole, felled by Mel Dale, mounted on a cradle for trucking to the pole yard in Enderby, 1977. EDMS 2887

Contracting is a tough business. It was tough from a company point of view, but tough for a contractor too. You're continually bickering with contractors about rates. Everybody thinks they got the worst timber, and they got the cleanup jobs, and you give this other guy this area where his timber's good and his ground is flat and he's making money, and you send me up the mountaintop somewhere. That goes on, that continuous hassle with them, but the individuality of those guys is really admirable. I kind of sit back and tease them, like one guy that just sold out. He sold a couple of million dollars' worth of iron and got rid of everything, and now he's down in Mexico all winter in a great big $400,000 motor home. That's the same guy that sat across the table from me, crying that he was gonna go broke tomorrow morning!

But some of them, well quite a few of them, didn't make it. That made it tough for us too because you like to establish a working relationship with the guys that are supplying your logs and keep them in business. It's easier than dealing with different guys every day. And you felt a responsibility for them too.

I never had a written contract with a logger until the early 1980s. There was a push on, they wanted the security. It was a difficult thing to develop a contract for them because the company had trouble guaranteeing that they were going to take x-number of loads every year because they didn't know what the market was going to be or anything else, and yet the guy wanted that information for security at the bank. But while the market was good, these guys didn't have a helluva lot of trouble getting financing.

I think the banks were tightening up and that's why the guys came to us for contracts. We finally developed a contract that they wanted. Not exactly what they wanted—they never got exactly what they wanted—but it gave them a modicum of security. There was a lot of weasling in it because you used to rate your contractors on how many loads a day he hauled. You call him a twenty-five-load-a-day contractor or a five-load-a-day contractor. And a contract has to say that it is dependent on other circumstances. You couldn't have a contract for five years that bound you to take twenty-five loads from this guy because you might not need them, so that was a difficult deal.

I guess to them it was quite a hassle getting money for equipment. Some of the ones that didn't make it couldn't handle the money. Most of them started out with very little and they ended up with millions of dollars' worth of iron on their hands. Some of them just weren't able to handle the business end of it. A guy that hasn't had anything, although he's a hard worker, if he doesn't pay attention or have a bit of a business head on him, the monthly cheques from the company if he's really producing are pretty substantial chunks of money. I think some of them just went hog-wild with it and eventually didn't pay their bills. They lived high off the hog while they were going broke. But the majority of them survived in our outfit. Only the odd one couldn't make it.

—Bill Uphill

Initially, because of their limited capital resources, there were few full-phase contract loggers. Instead there was a large number of single-phase contractors, each of whom owned one or two pieces of equipment that were usually operated by the owner. As time passed, the more successful of these contractors took on multi-phase contracts and evolved into full-phase operators. Many of them, like Roger Getz in Williams Lake, whose company, San Jose Logging, does contracts with Lignum, grew along with the quota-holding mills for which they logged.

The evolution of a contract logging sector, employed by the larger, more financially secure mills holding timber quotas, was not painless. Logging is a very risky business, particularly during periods of rapid technical change. There are no two identical trees in BC: however slight the differences, every single tree has to be logged individually. Performing this task under the wildly variable conditions of climate and terrain found in the Interior takes an enormous amount of skill. In recent times, doing the job within the regulations enforced by an employee of the Forest Service, who may never have logged in his or her life, requires the diplomatic skills of an ambassador. And doing it for the price most quota-holding companies are willing to pay takes a financial genius. It is no surprise that relations between the various players in this enterprise often were less than cordial.

> Some of the old-time fallers did hand falling all their lives. They got out in the bush when they were eighteen and they were in their forties when mechanization came. We had a number of them on our crew and they could not adapt to learning a machine. Particularly a feller-buncher, it's pretty complicated. Or a log processor. I always maintained that we should go into the nearest video arcade and hire kids there with the hand-eye co-ordination. But we had to deal with them. Sometimes we bridged their retirement and things like that. The payoff was that the new crew was smaller, was much more productive with this machinery. Of course we had the capital cost of the machinery, too. But the younger ones in our crews surprised me. Some of them that I never thought would adapt, even from running a rubber-tired skidder or bucking at the landing, moved up to a log processing machine. And the younger fallers, I can think of three of them that we trained minimally and, Jesus, they turned out to be dandy operators. Age, I think. You've gotta start out young on these machines. It's something that older people have trouble adapting to.
> —Bill Uphill

In spite of this, between the mid-1960s and the late 1970s, the logging and sawmilling sectors of the Interior forest industry developed rapidly to take their place among the most productive and efficient in North America, if not the world. In a little more than a decade the industry moved from its reliance on horses and bush mills to become highly sophisticated users of modern logging and milling technology.

By 1972 the industry was positioned to reap the benefits of its transformation. That year, lumber prices reached record highs and the biggest building boom ever swept across North America. The volume of wood logged in the Interior was three times the volume cut fifteen years earlier. Coastal production during this time increased by only 30 percent.

That same year, the first socialist democratic government in the province's history was elected. The forest industry was not sure what to expect, but because Premier Dave Barrett's New Democratic Party had some old policy on its books about nationalizing large resource companies, there was concern—particularly among those operators around Quesnel and Prince George who had once left Saskatchewan in opposition to social democratic policies there.

It turned out the NDP did not have any idea what to do with the forest industry. Once in office it was clear that promises to nationalize the industry had been rhetoric. Forest Minister Bob Williams, an academic urban planner whose quiet, shy demeanour was misread as cold aloofness, spent most of his first few months in office trying to learn from the Forest Service how the industry operated. His primary interest appeared to be the extraction of higher stumpage charges from the industry.

In its first year, with markets still booming, the government made its first significant forest

> When I come over [to the east Kootenays], the first few months I was on fires. They had a big fire in the Wild Horse the first year I was here. I worked on the fire and the forestry guy that was there, he kind of liked what I did, so ever after that he singled me out for fires. If there was a fire, I had to go.
>
> On a fire, if you got out there in time you just went with a hand crew. The night was the best time to fight fire because you could see and it wasn't going anyplace much. It moved very slowly at night. Even in '85 when we had the tremendous, horrendous fires, at night you could work right beside the fire. The daytime they took off, every day. Just an ungodly, blazing inferno. As high as you could see. And they burnt themselves out. I never saw anything like that in my life. I saw lots of fires but I never saw anything like that. They were all the same. We had six burning out of control at the same time in six different valleys. And they all reacted the same. Between eleven and one o'clock they would take off.
>
> Just burning. Usually it wasn't particularly windy that year. Just as high as you could see, just a mighty roar. When they went out, there wasn't a puff of smoke in where they burnt, they were absolutely out. The reason for that was it was zero humidity. It was so goddamn hot there was no oxygen. They burnt themselves out of oxygen. And they went out stone cold. But the problem was they spotted. Mile, mile and a half, two miles ahead. Hot coals. Once they got up into the atmosphere, they drifted. The ashes and cinders and that drifted, and that's what you had to go and get. The main fires burnt themselves out.
>
> [Before we had helicopters], if fires were spotted at night and you were called out, you would take a crew and position them thirty, forty feet apart with machetes and start digging right beside the fires. A one-foot fire guard, if you're right beside the fire, is as good as a mile-wide one if you're a little ways away. We'd build these huge fire guards. But that's all bullshit, because huge fire guards don't work if the fire gets a run at it. The old method of fighting fire right beside the fire is still the best method if you can get close to it.
>
> —Harry Meyers

sector move: it bought the Crown Zellerbach coastal pulp mill at Ocean Falls because, it said, jobs needed to be protected. The aging mill had run through most of its timber supply and Crown felt the cost of a much-needed upgrade could not be justified. Less than a month later, Williams announced the government was buying control of Canadian Cellulose, with pulp mills at Prince Rupert and Castlegar. The government explained that the company's New York owners were considering the sale of Celgar to another US firm, Weyerhaeuser. Williams argued that investment in the Prince Rupert operations was crucial to the development of the northwest Interior, the same rationale that was used twenty-five years later by another NDP government when it directed $400 million into the ailing Prince Rupert pulp mill.

By late 1973 the lumber market had begun to fall. The Timber Products Stabilization Act, which authorized the government to intervene heavily in the sale of forest products, was introduced. Most of the act never came into effect, but it did precipitate a dramatic increase in the price paid by Interior pulp mills for sawmill residues. The independent sawmill sector was mighty pleased with this outcome, but still wary of the NDP government.

In early 1974 the government intervened in the industry again, pressuring the Martens brothers to sell it their Plateau Mills operation at Vanderhoof. The Martenses were about to sell it to Rayonier, an American company, when Williams vetoed the transfer of cutting rights and matched the Rayonier offer. The brothers had no choice but to sell. A week later Williams bought Kootenay Forest Products, which had a mill in Nelson and extensive timber rights in the Lardeau region. Part of the reason for this move was to prevent a possible sale to Crestbrook, which had minority Japanese ownership.

> Some years we had to burn slash. The Forest Service had these regulations. You had to burn. And you had to burn a hundred percent. Not 99 percent. A hundred percent. There's sidehills with a southern exposure and very light soils, you got no business burning them. You burn the soil. All the time I worked in the forests I always argued with the Forest Service that you can't make up rules in Victoria, or anyplace else, and enforce those rules all across the province. It doesn't work that way. It's site specific. You must go out there on the site and make a decision what should be done there. Walk that log. Because our areas are so different. You can go out in the bush and every mile you've got different types of soil. More decisions made in the field. Because it's fine to print a big book and say this is how the logging's going to be done in BC, but it doesn't work that way.
> —Harry Meyers

By the end of the year, lumber markets had hit bottom and the province was wracked by strikes, including those staged by the pulp unions. On the Coast, as well as in the Interior, big integrated forest companies and smaller independent operators brought their conflicts and differences into the open. In the northern Interior a group of sawmills banded together and issued a statement that the pulp mills were underpaying them for chips. Early in 1975 the government acknowledged it had no coherent set of forest policy changes and appointed a Royal Commission to examine the industry. That fall it was defeated in a general election.

Peter Pearse, who had been appointed sole commissioner to conduct the inquiry, found himself in an unusual position: two months after his commission had been set up, the government which had appointed him was defeated. His report would be delivered to a new, unknown government, Bill Bennett's Social Credit administration.

Pearse grew up in the north Okanagan, the son of a pole maker. He studied forestry at the University of BC, where he achieved exceptionally high marks, and did graduate work in forest economics in Scotland. He returned to teach at UBC, and became a highly regarded academic who had grown up in, and had a firsthand knowledge of, the Interior forest industry.

In his report, published in 1976, Pearse covered the full range of policy issues, focussing on tenure policy, the means by which forest companies obtain government-owned timber. He raised questions about the number of informal regulations which had evolved in BC, particularly over the previous twenty years, including such provisions as quota. But his major concern was with the concentration of cutting rights into fewer and fewer companies:

> It is not the integration of the industry that is of concern here, but rather the policy-induced integration within individual firms. Nor, I want to emphasize, is my anxiety about the size, per se, of our large forest companies; it is the erosion of opportunities for others to play a constructive role in the industry and the growth of regional monopolies as large corporations assimilate small firms with their resource rights... the extent of industrial consolidation has proceeded well beyond what can be considered to have been necessary to keep pace with technological change and efficiencies of scale. I have taken the position in this report that in the absence of clear evidence that larger, more integrated corporations are substantially more efficient, forest policy should not be biased in their favour... In my opinion the continuing consolidation of the industry, and especially the rights to Crown timber, into a handful of large corporations is a matter of urgent public concern.

Elsewhere in his report Pearse argued that provincial forest policy should be adjusted to focus on protecting and enhancing the forests, not only to ensure the maintenance of timber production, but also to protect nontimber aspects. His general prescriptions were backed up by scores of specific recommendations. The report acknowledged growing demands for change from within and outside the industry; in itself, the report also helped bring about a dramatic shift in the BC forest industry over the next twenty years.

CHAPTER 11

New Challenges

The 1980s & 1990s

The past two decades have been turbulent ones for the BC forest industry. During this time the coastal industry, which had already reached maturity, went into decline; but the Interior forest sector, in spite of having to deal with many of the same challenges, thrived.

In 1978, responding to Peter Pearse's Royal Commission report of 1976, Bill Bennett's Social Credit government adopted a new Forest Act, in which there was no response to Pearse's recommendations concerning corporate concentration. A three-man committee had been set up to translate the report into legislation. It was headed by Mike Apsey, who was associated with the Council of Forest Industries (COFI), and included one other member connected to the council. COFI represented the province's major forest companies, most of them on the coast. Northwood, for example, was not a member. And while several Interior companies were members, they had relatively little clout. This committee never once consulted with Ray Williston, the primary architect of the existing system, and his offer of assistance was ignored.

So perhaps it is not surprising that the new Forest Act further entrenched the position of the big companies, and further encouraged corporate concentration. The terms of Tree Farm Licences were increased slightly and most other forms of cutting rights consolidated in a new tenure, Forest Licences. Large tenure holders were protected even further from having to compete for timber. In a gesture to smaller companies, a Small Business Forest Enterprise Program was created, guaranteeing a portion of the timber supply to loggers operating independently of the mills. Timber in this program would be sold at auction. This program included a Woodlot Licence, permitting individuals, small firms and others to obtain small parcels of Crown land for forest management purposes.

After the new Forest Act was passed, Apsey was appointed Deputy Minister to oversee its application. One of his first moves was to undertake a massive reorganization of the Forest Service. The venerable ranger system of administration was abolished. It had functioned since the Forest Service was formed: forest districts had been divided into

Ranger Districts, each of which was administered by a ranger who was authorized to manage his district and who was accountable for everything which occurred within it. The system had its problems, but basically it worked well. It decentralized administrative power effectively, as rangers worked closely with companies and communities. In the new system, which was highly centralized, each of six regions was divided into several districts. All personnel were centralized in district offices and given new functional titles—Resource Officer, District Manager, and so on. Most of the senior jobs were given to staff with academic credentials, while most of the rangers, the Forest Service's most experienced and knowledgeable personnel, were relegated to lower positions.

> I never believed in changing legislation. My reason for not changing legislation is that I could rule with regulation, and if the thing was wrong I could change it the next week with a regulation in the Cabinet and not wait until I had to argue it out in the legislature and get all the lawyers and everybody else in there. So Peter Pearse rightly said I hadn't done anything with basically changing the Forest Act. I accept that. I didn't, because I was working on regulations most of the time. He devised a structure under which the Forest Act would rule the industry. And remember, he just got his report finished when the NDP went out and Tom Waterland came in. Tom was a mining engineer and he had no background in forestry. So Tom reached out and he got Mike Apsey and Bob Wood. Johnny Stokes was there but Mike Apsey was the chairman of the committee that took the Royal Commission report and adapted it. Mike had come out from the Council of Forest Industries and Bob had come out from the Council of Forest Industries and the two of them enacted this change. I'd come back from New Brunswick by this time and you couldn't even talk to them. They didn't want to hear one thing, and nobody wanted to do anything any different than they'd been doing it before.
> —Ray Williston

The Forest Service was thrown into chaos. In 1983, two or three years into the reorganization and at a time of tight budgets, the government realized the new system was costing a fortune. Completion of the reorganization was suspended, and hundreds of Forest Service personnel were laid off as part of a "restraint program." This kind of about-face would have caused the industry a great deal of grief even at the best of times, but there were several external factors applying pressure to the industry as well.

Markets for BC forest products had begun to fluctuate wildly. In 1979 prices and sales reached record highs. The following year they plunged dramatically, following the rest of the North American economy into a prolonged recession that was harder on the industry than any downturn since the Great Depression of the 1930s. After about five years of reduced market strength, demand and prices recovered and rose to new heights in the mid-1990s, but then they tumbled again.

> I started trucking out of high school, and then in 1971 another fellow and I bought out a guy with five trucks. We come up with $5,000 each and that was our down payment to this fellow for five logging trucks. He was logging and he wanted out of the trucking end of it. The fellow I went partners with was a mechanic for the outfit so for five grand apiece we were the proud owners of five logging trucks and a contract to truck for this fellow. As we trucked for him, he'd take so much each month for the payments on it, and eventually we had 'em all paid for. We went as far as 1981 with a partnership and then I bought my partner out. Then I got the ultimatum that I had to buy the logging outfit or take my chances on losing the job with all my trucks too. I had to buy the other half of the outfit, and I ended up with the logging and the trucking.
> —Dave Webster, the owner of Deejay Trucking at Hazelton.

By the mid-1980s the Interior industry had completed an intense modernization phase, emerging with the most sophisticated and efficient logging and sawmilling sectors on the

Page 211: Shelterwood logging with an Owren cable yarder at Canadian Forest Products' Chetwynd operation, 1996. CFP

Page 212: A typical work station in a modern sawmill, this one at a Canadian Forest Products mill in its northern operations. CFP

A Valmet forwarder decks logs at Canadian Forest Products' Fort St. John logging site. These logs have been delimbed and cut to length at the stump, self-loaded onto the forwarder and hauled to roadside, where they will be loaded on a truck. Known as the Scandinavian system, this method of logging is rapidly gaining favour in the Interior because of its low impact on the forest. CFP

continent. Even as prices fell and producers elsewhere on the continent went out of business, the Interior industry was able to keep cutting lumber, shipping more and more of it into the USA. In 1982 the US industry responded by seeking a tariff on Canadian lumber. It lost, but returned immediately to try again. This attempt, which concluded in 1986, was more successful. The BC government responded with increased stumpage rates to forestall a US tariff. Over the next ten years, the Canada–US lumber tariff dispute became institutionalized, particularly for Interior mills. In 1997 every mill selling into US markets was given a sales quota, and any excess was subject to high tariffs. The policy has locked the industry into a costly and inflexible market restraint.

Another factor affecting the industry has been a growing public objection to the industry's use of the forests, which began in the late 1970s in tandem with an increasing global awareness of the impact human activities have on natural environments. Peter Pearse had noted these changing attitudes in his report. The Forest Act in 1978 did respond to this part of Pearse's report. It redefined the Forest Service's purposes and functions to include the care of fish, wildlife, water, recreation and other resources—as well as timber. The Forest Service was now required to integrate the management of these resources in consultation and co-operation

Above: Loading tree-length logs with a hydraulic loader at Canadian Forest Products' Takla operation. CFP

Left: Falling with a Timberjack tracked feller-buncher at Canadian Forest Products' Polar division, 1996. CFP

with other government agencies and the public. The Forest Service, which had already been severely disrupted during its reorganization in the early 1980s, was in no condition to fulfill this new function, let alone respond satisfactorily to the demands of impassioned nontimber interest groups.

The brunt of the environmental conflict was felt by the coastal industry, but the Interior did not escape unscathed. The first major fight was over an area in the northern Purcell Mountains. Under pressure from a broad coalition of outdoor organizations, the government created a wilderness reserve where logging was prohibited. Before long, other groups lobbied to protect the Valhalla mountain range in the upper Slocan Valley. This campaign succeeded also, and out of it emerged the Valhalla Wilderness Society, one of several environmental organizations throughout the province which pressured the government and engaged the forest industry in a series of disputes over particular logging sites. The Forest Service was unable to mediate, let alone resolve, these conflicts. Like the fight with US lumber producers,

the conflict over forest use became institutionalized; at the close of the century it remains an unresolved and costly factor for the industry.

During the late 1980s, Dave Parker, the first professional forester to serve as Forest Minister, touched off a furor when he announced his intention to convert BC Forest Products' Forest Licence at Mackenzie into a Tree Farm Licence. Parker hoped to convert most Forest Licences in a similar manner, and to drum up support he toured the province and appeared at a series of public meetings. He met a solid wall of resistance, including a great deal from within the forest industry. At his last meeting, on Vancouver Island, the hall in which he spoke was surrounded by hundreds of forest workers, demanding another Royal Commission on the forest industry.

Instead, the Bill Vander Zalm Social Credit government set up a "permanent" Forest Resources Commission (FRC) to provide forest policy advice on a continuing basis. The FRC held a series of public hearings around the province and received more than two thousand formal submissions—almost ten times the number presented to Pearse. Its report, issued in 1991, called for "enhanced stewardship" of forests and made some specific recommendations, the most notable being to involve the public in forest management planning and to adopt a forest practices code. The report also proposed to change the forest tenure system radically by creating a Crown corporation that would be given title to all commercial forest land in the province. This proposal, which would create yet another level of quasi-government bureaucracy, was quickly and quietly shelved; within a couple of years the "permanent" FRC itself disappeared.

A few months after the report appeared, a New Democratic Party government was elected under the leadership of Mike Harcourt, and attempted to deal head-on with the growing conflict between the forest industry and nontimber forest interests. The government set up a permanent Commission on Resources and the Environment (CORE) to develop land use plans for several regions, including the Cariboo–Chilcotin, East Kootenay and West Kootenay–Boundary in the Interior. In response to a growing threat of consumer boycotts in the United States and Europe, the Harcourt government pushed ahead with a new forest practices code, which had been partially completed by the previous government. In addition, the government increased substantially the number and size of parks and protected areas.

The consequences to industry of these growing public pressures, and the government's attempts to deal with them, were considerable. Through the mid-1990s, the industry operated in a constant state of crisis. Public protest increased and the expiry of the Memorandum of Understanding on softwood lumber exports to the USA heralded a new round of charges that Canadian exports were being subsidized by low stumpage. The CORE process appeared to be interminable and inconclusive; eventually the commission was shut down. In 1995 the Ministry of Forests unveiled the Forest Practices Code, which consisted of a vast body of detailed regulations backed up by million-dollar fines and a large bureaucratic enforcement agency. Nevertheless, the code was supported by many industry leaders as a necessary response

[The reduced availability of high-quality timber has had an effect on plywood production.] No question, because if you're selling to Japan, a nice clear face, they like that. High-quality wood is limited. The other thing that's hurt is that you can produce out of that same peeler log a two-by-ten much more effectively and get a better margin than you could making a sheet of plywood. At Quesnel, which is one of our best wood supplies, we can make some beautiful plywood yet, but the cost of wood is so high we're going to say, if the appearance is not important, we're going to make it out of a lower-grade log rather than the higher-grade log.

I went down to Seattle about two months ago. Our parent company, Champion, has some land there. They were sorting logs and one peeler sort was five to eight inches, top diameter. Peeler logs? Now what the hell are you talking about? I couldn't believe it. What it is, is the cost of peelers is so high these guys can use the inside stuff for cores. You know, the Interior's looking at aspen. We're great innovators when we get our butts to the wall. That's what's happening.

But on the other hand, plywood for certain uses, for certain things, is much better than OSB [oriented strand board]. You have to make a much thicker panel to make the same thing of OSB. Also there's moisture. There will always be a use for plywood. The question you have to face is how the hell can you compete with some of these big plants? On a three-eighths-inch basis, say Quesnel does 180 to 200 million feet a year. Some of those big plants are doing 300 and 400, and we have three hundred people and they have one hundred. So on a cost basis, you're really in trouble, unless you've got certain markets you can satisfy. And of course the other thing you've got going for you is if you've got the plant built you don't have to put a hundred million in, you don't have to write it off. But if you looked ahead, there's gonna be a lot of the cladding, sheeting—things will be done with reconstituted board. It just makes sense. It has a disadvantage to carpenters, in that it's relatively heavy. The answer may be that they'll tack the roofs together and put them up with a crane, so it won't make any bloody difference.

—*George Richards, president of Weldwood Canada. He graduated in forestry from University of BC in 1959.*

Delimbing and bucking whole-tree logs at Slocan Forest Products' Quesnel operation, 1995. This is a fully mechanized show, in which a feller-buncher is used to fall the trees, a grapple skidder to move them to the landing and a processor to delimb and cut them to length. SG

Right: A John Deere grapple skidder is used to skid a turn of whole-tree logs to roadside at a Canadian Forest Products operation near Fort St. James, 1980. CFP

Below: Grapple yarding bundles which have been felled with a feller-buncher at one of Canadian Forest Products' Chetwynd shows. The machine in the rear acts as a tail-hold for the grapple yarder's cables and moves along the back of the setting as the yarder progresses. CFP

to public concerns at the time, and it did have the effect of easing tension between industry and its critics. After two years of operation, however, it was found to have increased logging costs by more than $12 a cubic metre. Costs were increased even more by the creation of another government agency, Forest Renewal BC, paid for by increased stumpage charges. The increases were a response to the US softwood dispute and FRBC was conceived as a way of returning the tax revenues to the industry in various ways intended to strengthen it, although some critics feel it was used to find work for forest workers displaced by the government's other forest sector initiatives. All of these events took place against a background of wildly fluctuating markets and the ongoing tariff dispute with US lumber producers.

Loading a Trans Gesco clam-bunk skidder near Quesnel at a Slocan Forest Products operation. SG

In spite of this twenty-year period of disorder, the Interior industry continued to grow, largely through a combined process of consolidation and technical innovation. Since 1978 a few Interior companies have grown, partially by the acquisition of others and partially by adopting new processing methods and product lines. But the consolidation process in the Interior has been distinctly different from that on the Coast, where the industry increasingly came under the control of large, integrated multinational corporations. In the Interior, consolidation has tended to enable small local firms to grow into large regional or provincial entities.

Throughout the 1970s, for instance, Sam Ketcham used the new Chip-N-Saws to create West Fraser Timber. After establishing the first mill at Quesnel, he bought up a series of small mills—first at Williams Lake, then at Smithers, Fraser Lake and Dawson Creek—that were not equipped to cut the timber profile available through their quotas. He installed the new saws and turned each acquisition into a profitable enterprise. It was claimed at the time that a new Chip-N-Saw was worth a million dollars in profits, and Ketcham proved it several times over. By the time he died in a helicopter crash in 1977, West Fraser was in a position to buy larger operations, such as Chetwynd Forest Industries, which it took over in 1979.

At this stage in its development, West Fraser was producing a lot of chips and found itself at the mercy of the pulp mills, which set chip prices. A deal was struck with Daishowa of Japan to build a pulp mill that could utilize chips produced in Quesnel, and a share of the Eurocan pulp mill at Kitimat was bought to make use of chips coming from the Smithers and Fraser Lake sawmills. Now established on the Coast, in 1984 West Fraser bought one of the big Terrace sawmills which had been built to cut lumber from Tree Farm Licence #1 timber belonging to the Prince Rupert pulp and paper-based operations, by then owned by Westar. Another coastal mill was built at Prince Rupert a few years later. In 1988 West Fraser bought the chain of retail stores developed by Alberta-based Revelstoke Sawmills.

In the Okanagan another family company, Tolko, went through a somewhat different expansion phase. The Thorlaksons built up their Lavington-based operations through the 1970s, which included the purchase of the Dumont family's sawmill at Newport Creek. When it burned, its timber supply and workforce were redirected to Lavington, which doubled the output there. In 1981, just five months before Harold Thorlakson died, Tolko acquired John Ernst's Quesnel sawmill, doubling production again. Balco Industries, the Heffley-based sawmill started by the Balison family, was picked up in 1987.

NEW CHALLENGES • THE 1980s & 1990s

> I no sooner got back here to the Okanagan in '67 than I saw the very significant discrepancy between the forest industry in the southern part of the province and what I'd seen up in the Cariboo. They hadn't adjusted to small-log utilization down here, and in the Cariboo they were into lodgepole pine and small-log utilization. Down here all the operators were still in intermediate utilization and in many ways resisting the trend. Ray Williston adopted the policy that if you had barkers and chippers you could get a small-log, third-band timber allocation based on performance. We technically didn't have a barker and chipper in our mill but we had a fuel wood system in the portable mill that we sold the sawdust for fuel. We sold all the slabs for fuel. And of course we sold the lumber. So we had what we claimed was better utilization than the mills that had barkers and chippers. I sent a brief to Ray Williston, went to Victoria and met with him, and he said, You qualify but you gotta put in a barker and chipper. We'll let you qualify once you've got your barker and chipper in. So we took the old portable mill and set it up on a foundation, a concrete foundation, and put a barker and chipper in and we got a very very small, modest third-band allocation. We've grown from there.
>
> You have to remember we're still on the Interior plateau here. The Okanagan is just a valley within the Interior plateau and it runs and bumps into the Monashee Mountains. So the lodgepole pine opportunities are basically in the plateau area, and as you run into the Monashees you get into more a wet belt timber type so you get more spruce, some of the cedar types. The cedar types are in the transition zones as you get into the wetter areas on the foot of the Monashees and the Shuswap area. One challenge in this area is to deal with all the species. We have seven or nine species, and that's one of the things that really forced us in the southern Interior to be much more market focussed, more flexible operators than mills in the north. Fellows up north basically use one species. They got very good, very efficient and were very successful as commodity operators. Down in the south here you had to be flexible, you had to be good at a lot more things. We couldn't build the cookie-cutter mills that the guys in the Cariboo were building, or in around Prince George. Which was a reason for the demise of a lot of the operators in the south. They couldn't adapt. You had to figure out where you were making a buck and where you were losing a buck and do what you did best. Lots of times you could make money on one species and then you'd lose it all on the other species. To make a buck in the south was a lot harder than in the north.
> —Al Thorlakson, president and CEO of Tolko Industries.

Balco itself had gone through a major ownership cycle during the previous decade. By 1970 the company was being run by the second generation of Balison sawmillers, and they needed capital to put in a plywood plant. A deal was struck with Peter Bentley and in 1973 Canfor took 51 percent of Balco shares, with Dave Balison staying on as president. Under this arrangement Balco bought out Ken Long's mills, one at Louis Creek and the Nicola Valley Pine Mills at Merritt. Balison retired in 1979. In 1986 Canfor sold its interest in Balco to a diverse group, and control returned to the Balison family. Tolko took over in 1987, again doubling its production capability.

In 1988 Tolko began to expand beyond the Interior and bought the Nova Lumber cedar mill in North Vancouver. In 1992 the retail building supply business begun by Harold Thorlakson, which formed the basis of the family business, was sold. Sensing problems ahead for the BC forest industry, the company diversified by expanding outside the province, starting with an oriented strand board plant at High Prairie, Alberta, in 1995. In 1997, Tolko shut down its coastal operations, and later that year it bought the troubled Montreal-based Repap's Manitoba pulp and paper mill and sawmill, all located at The Pas.

The Ainsworth family went through the same kind of expansion phase from their base at 100 Mile House, under the name Ainsworth Lumber. In the late 1960s they added a dry kiln and chipping and debarking equipment, and began shipping chips to Prince George. In 1970 they built their sawmill complex at Exeter, on the Pacific Great Eastern Railway just west of 100 Mile House. Four years later they built a second mill of about the same capacity at

Chasm, the site of the old Clinton Sawmill planing mill. The company diversified somewhat in 1975 when it acquired Exco Industries, a 100 Mile House manufacturer of sawmill and logging equipment.

Ainsworth then decided to focus on high-quality studs, and added a finger-jointing mill to the Exeter operation in 1978. It was moved to Abbotsford in 1986. The following year Ainsworth bought Evans Forest Products' operations in Savona and Lillooet, which included two sawmills, a veneer mill in Lillooet and a specialty plywood plant at Savona. In the mid-1990s two oriented strand board plants were built, one at 100 Mile House and the other at Grand Prairie, Alberta.

In '69 we put in the barkers and chippers at Lavington and we got our first third-band allocation. I think it was in 1970 we started up our first small dry kiln. I remember vividly, it was a memorable event. We started that dry kiln up and the price of kiln-dried SPF lumber went from $96 to $104. We thought we'd died and gone to heaven! Just as we started up our dry kiln the price of dry lumber went up and we were really thankful we hit it right.

In '81 we bought Ernst Sawmills in Quesnel. It couldn't have been a worse time in many ways. The market was in the dumps and John Ernst, he was a character. I talked to John, and we agreed to disagree that his price expectations and our value was out of line. He then tried to sell to another guy. That fell through and then we showed up again and sort of picked it up, in a bridesmaid kind of deal. After we closed the deal on an interim agreement, John wanted to renegotiate. He thought this was but a stepping stone to a new agreement, new price. His expectations went up and we said, No, John, we have a basic agreement. He says, Oh no, we've got to do a final agreement. We said, Sure, we'll do a final agreement which reflects our interim agreement. He says, Oh, no, no. I want a different price.

We went around and around and around on this thing. Then we took a look at our interim agreement and said, Look, we can close on our interim agreement. It's a good agreement. Everything's there that needs to be there. Our lawyers had done a good interim agreement so we said, Okay, we want to close, and John said, You can't close. He wouldn't even come to a meeting at this stage. Once he realized we wanted to close on the interim agreement he just refused to show up. We were in a quandary as to what to do, so we brought in some other lawyers to take a look at our agreement. This one character said, Why don't we tender our offer? We said, What good will that do? He says, Well, you just wait and see.

We phoned up John's lawyer and said, We're gonna tender our offer, deliver our purchase price, we're gonna bring our offer to his office. His lawyer kind of yawned until our lawyer informed him we're gonna bring it in cash. And then his lawyer really got uptight to think that we're gonna deliver millions of dollars. We asked him what denomination did he want the bills. Well, I can assure you, John Ernst showed up at a meeting the next day.
—Al Thorlakson

Riverside, another family-owned sawmill company with bush mill origins, grew in a dramatically different manner. After acquiring several small firms in the north Okanagan in the 1960s, the company spent the next twenty-five years growing to become one of the Interior's more efficient and profitable firms. In 1993 it bought Fletcher Challenge's Interior operations. This was the successor company to Crown Zellerbach, the multinational coastal company which had developed a large Interior presence, beginning with the purchase of S.M. Simpson at Kelowna in 1965. Crown took over Armstrong Sawmill's plants at Armstrong, Enderby, Vernon and Falkland in 1969, and Ponderosa Pine Lumber at Monte Lake the following year. In the late 1970s these operations were consolidated into two mills at Kelowna and Armstrong, which were sold to Riverside in 1993. A year later, Riverside bought Jacobson Brothers Forest Products' sawmill in Williams Lake.

Most of the Interior's integrated companies also underwent some degree of expansion after 1978. Canfor, which had taken over Balco for a few years and then cut it loose, bought

A self-loading logging truck putting on a load at a Slocan Forest Products show in the Slocan Valley. SG

Balfour-Guthrie in 1989, acquiring four sawmills in the process—the Netherlands Overseas mill in Prince George, and others at Taylor, Bear Lake and Clear Lake. The following year it sold its mill in High Level, Alberta to Daishowa Canada. In 1995 Canfor's purchase of a share in two Prince George companies enmeshed it deeply into the Byzantine web of industrial ownership in the north.

These companies, Lakeland Mills and The Pas, both had deep roots in the Prince George lumber business, and were both owned by three local businessmen. George Killy, the youngest of the trio, had learned the trade from his father, Ivor, who had been involved with several local companies since he started out cutting ties in the 1930s. The other two, Ivan Andersen and Bob Stewart, are an inseparable pair of characters, now well into their eighties and still extremely active in the business.

Andersen and Stewart own Sinclar Enterprises, a wholesale lumber firm whose tentacles reach into various northern BC forest sector enterprises. Andersen started working at Sinclair Mills east of Prince George in the 1930s, and Stewart began working there during the war. With the decline of the East Line mills, the two set up the first wholesale lumber business in the northern Interior in 1962. From there they began to acquire an interest in various mills, and in 1973, with George Killy, they bought Lakeland. The three Lakeland partners bought Helco Forest Products in 1978; in 1987, when the Winton family decided to get out of the forest industry, they bought The Pas. In 1995 Killy sold his one-third share in these companies to Canfor, thereby creating the improbable, but apparently successful, alliance of Peter Bentley, Ivan Andersen and Bob Stewart.

Following pages: The Pas Lumber's planer mill in Prince George, 1982. TPL

The other integrated Prince George company, Northwood, went through a similar type of expansion, buying Rustad Brothers' Prince George mill in 1991. Since its initial sawmill buying spree when it was set up in the 1960s, Northwood had consolidated its original purchases and picked up another substantial sawmill company, Bulkley Valley Forest Industries, which had a mill at Houston. In 1980 it bought two other Prince George companies with interwoven ownerships. Prince George Wood Preserving was set up in 1977 by Gordon Swanky and Don Flynn to treat lumber. In order to maintain its wood supply, it got into sawmilling and quickly evolved into a major primary producer. A decade or so after it was founded, the company was sold to Prince George-area loggers Chris and Paul Winther. These men were also partners, along with Swanky, Flynn, Howard Lloyd and several others, in the city's only plywood plant, North Central Plywood, established about 1970. Northwood bought North Central as well.

After it was formed in 1964, incorporating John Bene's Western Plywood, Weldwood also pursued a course of diversified expansion. Right after it was formed it acquired an extensive coastal operation, and in 1969 it built Cariboo Pulp & Paper with Daishowa at Quesnel. In the mid-1970s it entered into two joint venture sawmills in the northern Interior, Babine Forest Products at Burns Lake and Houston Forest Products at Houston. In the late 1980s it acquired Champion International's pulp and sawmill complex at Hinton, Alberta, in exchange for giving Champion control of the company. In the mid-1990s it disposed of its coastal operations to focus its attention on the Interior.

In the southeast, at Cranbrook, Crestbrook went through the same kind of expansion process as its northern counterparts. Through the early 1970s it focussed on consolidating and modernizing its existing sawmills, and in 1983 the company bought its old rival, Crows Nest Industries (formerly Crow's Nest Lumber). This old coal mining company had become serious about lumber production in the 1960s and picked up several south Kootenay mills

Unloading a truck at Slocan Forest Products' Quesnel mill yard. SG

which were all consolidated in a new mill at Elko, opened in 1969. Its purchase boosted Crestbrook's timber supply and sawmill capacity by 50 percent. In the late 1980s, again backed by Japanese money, Crestbrook built a pulp and paper mill in northeastern Alberta.

The most spectacular example of corporate growth in the Interior forest industry during the past twenty years has been the formation and expansion of Slocan Forest Products. Slocan had its origins several decades earlier in the Slocan Valley, with the Passmore Lumber Company. In the 1960s it was bought by Axel Erickson, the pioneer helicopter logger from California. In 1969 Erickson sold the mill to the Meltzer family, a New York family of lumber merchants that wanted to establish their own supply source in BC. In addition to the Passmore Lumber mill, they bought another sawmill in New Westminster and a stud mill in Quesnel, named the whole operation Triangle Pacific Forest Products and installed Ian Mahood as vice president to run it. Mahood encountered problems with mill operations and called in Ike Barber, who had just finished his vice presidential stint at Eurocan. Mahood eventually left the company and Barber replaced him.

> The Meltzer family out of New York were very typical lumber merchants working off the Brooklyn docks, buying lumber from the boats and selling it up and down the Atlantic coast. They grew on that and were fairly significant people who would come around to Vancouver and buy green lumber by the boatload from the mills here on the coast, trundle it around to the docks up and down the Atlantic seaboard, land it there, then sell it off the docks. Then Abe Meltzer had the brilliant idea, Why don't I own some sawmills in British Columbia and then I won't have to hassle with these coastal people? I'll just bring lumber from my sawmills down to my docks and sell, and I will have vertical integration. Well, he didn't use those words, but that's what he was thinking about.
>
> So he bought three sawmills in BC. The first was the Pacific Pine Company in New Westminster. Then he bought the Slocan mill, and then he bought what was called the Tripac Stud Mill in Quesnel. So now he had three sawmills. Then he began to find out that these sawmills could sell the lumber to somebody else at higher prices than he wanted to pay and he could continue to buy his lumber from other mills at lower prices and feed it through his system. I don't think he ever sold very much lumber vertically up and down, which was his original idea.
>
> Then Ian Mahood ended up running the BC operations. None of the mills were doing very well, but the Quesnel mill in particular had been built as a stud mill. The Meltzer people bought it and decided they knew how to do it better so they tore the mill all apart, installed a stud head rig—which obviously required big logs—and they were going to really do well. Then they turned to their forester and said, Bring us the logs. So the forester brings in small, baby pine logs and the Meltzers say, No, no, no. Take these away. We want the big logs for which we built the mill. And the forester says, I'm sorry, Mr. Meltzer. That's all the logs that we have.
>
> There was a great furor, but that was very, very typical of the industry then—somewhat typical of the industry today—where engineering people and visionary people build the sawmill for their vision or what they perceive is an excellent engineered sawmill without specific reference to the forest base that feeds it.
>
> They were in a helluva jam. Ian says to me, "Why don't you go up there and help me sort this out?" So I went to Quesnel, and all we really did is draw a red line around the timber base that was gong to be attributable and assigned to the Tripac mill and do an engineering-forestry survey to define what that timber base could supply in terms of the range of log diameters and the volume of lineals that you would have to put through the mill. Then we did just a simple calculation of the lineals that you'd have to get through and the speeds that you would have to get. We rebuilt the mill yet one more time, threw out the big head rig and designed it for a small, high-speed throughput machine.
>
> Then Ian Mahood says, Why don't you go down to Slocan, I got problems down there too. So I went down to Slocan and there was another classic screwup, if you will, between the forestry operation and the manufacturing operation. They had installed a conventional sawmill. Then they installed a recovery mill which was designed to put through pulp grade logs, extract out whatever
>
> *Continued on next page*

Continued from previous page

number you could get and turn the rest into chips. Great theory, but it didn't work. They were logging pulp grade logs like crazy and trying to get them through this recovery mill, which had a major bottleneck. The Castlegar pulp mill would not buy the chips. So what they were doing is logging these pulp grade logs, putting 'em through the sawmill, the recovery mill, extracting out basically economy and utility lumber and burning the chips. All the time the forester was pushing pulp grade logs in that lake. So when I got there we had a lake full of pulp grade logs with nowhere to go. The focus was on that and off the sawmill. We just applied simple, ordinary thinking. When Ian left, I became the BC head of the Triangle Pacific operations.

Then Meltzers decided they wanted to sell. They phoned me up and said, We've sold the coastal operation to Ches Johnson at Whonnock. I said to them, Look, if you're going to do all this I don't want to preside over a wake. Why don't you see if I or we, the management group, can buy. So now we're talking about the Slocan mill.

Abe says, Well Jesus, Ike, you don't have any money to do this. How can you do this? What I didn't know is that he had probably tried to sell Slocan to everybody else in BC and no one would buy it. That's probably why he was very hospitable to my proposal. What I learned when you work with maybe any kind of people but particularly Jewish people, and these were Russian Jewish people, and I do not say this in any derogatory sense, but they either love you or hate you, one or the other and very little in between. I guess luckily enough for me, the relationship came down on the side of "we love you" and on the fact that he wanted out and no one else would buy it.

So he says, "Okay, this is what we'll do. We'll lend you some money and you go and make a feasibility study." He loaned us $70,000 and we hired Pemberton's to do a feasibility study. They came up with a combination of the Royal Bank and the British Columbia Development Corporation. Abe brought us all down to Texas, myself and the Pemberton people. I remember driving out in the taxi from the hotel to Abe's office and saying, "Look, guys, what are we gonna offer this man? How much are we gonna pay for this?" Pretty unsophisticated. We were in a taxi and each of us wrote a number on a piece of paper. We compared our numbers and we arrived at an offer. We ended up buying it for book value. Got the timber all for nothing.

We didn't have any money. This required four million of debt and two million of equity. Again Abe comes forward and lends us about $70,000 to hire Pemberton's to raise the equity. What a farce that was! Looking back, they didn't do a goddamn thing.

I approach my wife and say, Look, we really need to raise some equity to buy this sawmill. What would you say if we mortgaged our house? So anyway, we go down to the Royal Bank to talk about mortgaging and they say Fine, but your wife has to get legal advice on this. So we trundle over to see a lawyer and the lawyer says, I'm sorry, Mr. Barber, but you have to wait outside. I have to give your wife independent legal advice, on her own. So I sat in the anteroom while he talked to her. He fundamentally says, If you don't make the payments, the bank'll take your house. So she came out and handed me the mortgage and says, I hope you birds know what you're doing.

Ron Price, our chief financial officer, did something of the same thing. The third partner, Harry Argital, who grew up in the Slocan Valley and was the mill manager, wanted very much to be a part of this. He and his wife had a lovely house and acreage in the Slocan Valley, and his wife says, There's no bloody way you're gonna mortgage this. But he scraped up all he had. I took a trip to Toronto to talk to a venture capital outfit. I told 'em my story and they loaned us $500,000. So I think we came up with $900,000 out of the two million we needed.

And here's Abe Meltzer, coming up from New York to close this deal. I met him in the Vancouver Hotel, at night in the lobby, and I said, Abe, it doesn't look like we can do this deal 'cause all we can raise is $900,000 of equity. God, he sat me down in the lobby of the Vancouver Hotel and he just bawled the hell out of me for about an hour. Basically, about Trudeau and Canadians not having the guts to invest in their own country, and all this stuff. Anyway, the net effect, to make the deal, Abe left in a $1.2 million debenture at a fixed interest rate of 10 percent.

—*Ike Barber, president of the Slocan Group. He graduated in forestry from University of BC in 1950.*

When the Meltzers decided to pull out of BC in 1978, Barber and another Triangle Pacific manager, Ron Price, offered to buy the Slocan mill. They had no money, apart from what they could raise by talking their wives into remortgaging their houses, but managed to swing a deal to buy the mill and manage the Quesnel stud mill, under the name Slocan Forest Products. The next year they bought the Quesnel operation. At about this point, the lumber market collapsed. Somehow Barber and Price survived and attained a sufficiently sound bottom line to look seriously at another offer that came their way in 1985. Revelstoke Lumber wanted out of its east Kootenay mill, the old Wilder sawmill at Radium, so Slocan bought it.

In 1978 Abe Meltzer wanted to sell the Quesnel mill. We couldn't buy this for book value 'cause he had a lot of sentimental attachment to that mill. It had big inventories, and all this, and he wasn't gonna sell this mill at book value. So we were back and forth on price. I remember vividly that on Boxing Day I had all my family, my grandchildren and all, at our house. I get a phone call about 6:00 our time Boxing Day morning and it's Abe's secretary. She says, Mr. Barber, Abe wants to speak with you. He says, Ike, are you gonna buy this goddamn mill or aren't you?

"Yeah, Abe."

"Well, what about the price?"

"We're looking at this, we're looking at that."

"Look, you've looked at this bloody mill long enough, we're gonna settle a price right now."

Here I am at 6:00 in the bloody morning, phone in my ear, grandchildren running all over. So I blurted out some number and that was the price. He says, "Okay. I suppose you're gonna want another debenture."

"Yeah, Abe, I guess we are." I think he left in $900,000 on that one too.

—Ike Barber

A hauling contractor headed for the mill yard with a good load at Canadian Forest Products' Chetwynd division.
CFP

Milled lumber emerging from a modern gang saw at Canadian Forest Products' Isle Pierre sawmill west of Prince George. CFP

Two years later Slocan bought Clearwater Timber, a company established in 1940 by the Calgary-based Swanson family which owned sawmills in Clearwater, Vavenby and Valemont. Later that year another mill at Quesnel was purchased. To that point, all the operations had good timber supplies with mills incapable of cutting the timber profile. Barber's strategy was simple: rebuild the mill so it could cut the timber available efficiently. It worked, and by the time the market picked up in the late 1980s, they were all making money.

Then Slocan bought Tackama Forest Products at Fort Nelson. It consisted of timber, logging equipment, a sawmill, a plywood plant and a half interest, along with Riverside, in another plywood plant in Vancouver. Tackama was another joint venture launched in 1974 between the Fort Nelson Indian Band and a group of Prince George entrepreneurs, including Curt Garland, Bob Lunde and Lorne Goodlett. It was planned as a fully integrated complex, including a pulp mill. In the mid-1980s Tackama's local competition, Bob Cattermole's Fort Nelson Forest Industries, went broke. Tackama picked up its assets. For several years the partners attempted to raise the money for the planned pulp mill, but finally realized it was not going to happen and sold to Slocan in 1987. Slocan lined up the money for the pulp mill and even obtained a licence to run it, before deciding it was a bad idea.

In 1991 Slocan entered into a complex arrangement with Fibreco Export, of which it owned a share. Fibreco was set up in 1977 by a group of about twenty-five Interior sawmills seeking an alternative market for their chips, to get out from under the thumb of the Interior pulp mills that controlled prices. They formed a chip export company with a shipping terminal in North Vancouver. In the early 1990s a joint venture pulp mill at Taylor was built through Fibreco. Slocan acquired 80 percent of the venture and took on a contract to manage it.

So far Slocan's acquisitions had been initiated by vendors who had lost interest in the business and hoped Barber would take their companies off their hands. The next one to come along was the Plateau sawmill at Vanderhoof, but it was put out for bids. By this time Plateau, which had been squeezed out of the Martens family hands by the government, was held by Westar, the corporate entity which had evolved after the Bill Bennett Social Credit government dumped the NDP's industrial purchases back into the private sector. Now Westar was unloading Plateau, along with Celgar at Castlegar and the Prince Rupert pulp mill complex. Its Kootenay Forest Products mill at Nelson had been closed.

> It didn't make sense to me to be having chip trucks, both rail and truck, passing one another on the road going the wrong way. A lot of people were saying, We don't want to be taking chips from there. If that pulp mill can use 'em, why can't yours? You're all making softwood kraft. What's going on here? We'd better take a hard look at what we're doing. So we got together and we got a few trades going which were beneficial to all the companies really. Then a lot of the Interior chip suppliers began to feel pressure from pulp mills on pricing and not taking the volume, saying, we can't use your chips. Finally a bunch of them got together and formed their own company, Fibreco, and then they set up the shipments to Japan out of North Vancouver. That got the attention of the big pulp mills because they found out that they no longer could just say, Hey look, here's the price. You take it or swallow your chips. To protect their investment in their sawmills they needed to have a more secure outlet for chips. I don't think it hurt. It made everybody smarten up a little bit.
> —Fred Stinson

Westar poured a lot of money into the Plateau mill, and although it did not have an adequate timber supply secured, the mill was well suited to the available timber. There was some stiff competition from Prince George interests, but to their chagrin, Slocan won. A year later, in 1994, Barber was approached by the Royal Bank of Canada and offered Finlay Forest Industries' integrated complex at Mackenzie.

Launched by Bob Cattermole in the 1960s, Finlay had ended up with Fletcher Challenge, the successor to BC Forest Products, which owned the other integrated mill complex at Mackenzie. The operation was broke, and both Fletcher and the bank wanted out. Not particularly interested in getting into the paper business, Barber brought in Donohue, the US paper company. They split the deal, which was completed in 1993. In 1995, having concluded that a pulp mill at the Tackama mill in Fort Nelson was a poor idea, Slocan built an oriented strand board plant there instead.

> Finlay was something like the Skeena of today. It was founded by some mothership companies. They decided that they wanted to focus somewhere else. It was owned by the Royal Bank and Fletcher Challenge Canada. Fletcher gave the keys to the Royal Bank and said, We don't want anything more to do with it. So the Royal Bank was then stuck with a hundred percent of a timber base, two sawmills, an antiquated pulp mill and a newsprint mill. The antiquated pulp mill provided the pulp to furnish the newsprint mill. They struggled along under a management team for two or three years, and the bank says, "Hey, we want out."
> There was one other company looking at it, but fundamentally they came to us and said, Look, are you interested in this? Because we, as a bank, don't want to continue to manage it. We decided that, yeah, we were great on this but we weren't so great on newsprint, it sort of intimidated us. So who would be the best partner that knew the most about newsprint? We quickly decided on Donohue. The Donohue people had come out to look around BC and we had one of the investment houses arrange a lunch. We identified the issue to them. They picked up on it, so we created a fifty-fifty partnership. We've been at it four or five years now and it's sort of working. We made a significant investment of about $150 million to stabilize that operation. There were some startup problems, but it's a very good investment. Without us stepping in and investing that money, clearly it would have been another Skeena, 'cause they were basically broke and either didn't know it or didn't accept it, or some of both.
> —Ike Barber

After only a decade and a half, Slocan had grown from nothing to become the sixth largest forest company in BC, with sales of almost a billion dollars a year. Shares in the company were widely held. In February 1995 Canfor made an unsolicited offer for Slocan shares in an attempt to take control. The industry and the financial world were kept on the edge of their seats for several weeks while Ike Barber and Peter Bentley at Canfor competed publicly for shares. In the end Barber won, signing a big chip supply agreement with Weyerhaeuser and lining up the financial backing needed to provide shareholders with a more attractive offer.

Left: *An interior view of the Takla veneer plant at Fort St. James, later purchased by Canadian Forest Products and closed in 1982.* CFP

Below left: *Canadian Forest Products' Prince George Pulp & Paper mill, 1996.* CFP

If you drive down the road from Shelter Bay up the Cranberry River 'til you come out into the Columbia River, you drive through quite good-looking reproduction. That was all logged in the '60s and just left to natural regeneration. It wasn't planted at all. It looks pretty good now. What we went with was leaving seed blocks, more or less progressive clearcutting with ample seed blocks to reforest. Then we'd do our regeneration surveys, and if we had to plant we would do something about planting. But planting wasn't part of our initial plan at that time.

Regeneration is not really a big problem in the Interior. It comes back and it comes back right away. There's probably been a few acres here and there lost to deciduous but there's no vast areas of southern Interior forest land that have been lost. Anyway, you have to recognize that there wasn't a helluva lot of planting being done in the Interior at that time. We were doing what we thought we had to do, that's all. Basically relying on natural regeneration. What I've seen of it, going back, I'm not disappointed. I don't see that there's any criticism due. It was a system that we put in place and we stuck with that for twelve or fifteen years.

—*Fred Waldie, who graduated in forestry from the University of BC in 1952 and worked in his family's sawmill business at Castlegar until it was sold to Celgar. He was a senior manager at Celgar and its successor companies until 1982. He retired as wood supply manager at Cariboo Pulp in 1994.*

Insofar as the consolidation binge of the 1980s and early 1990s involved logging rights on Crown timber, it was precisely what Peter Pearse had warned about in his 1976 Royal Commission report. These same warnings had been repeated in the Forest Resource Commission's 1991 report, which pointed out that in 1975 the ten largest companies controlled 59 percent of the province's harvesting rights, and in 1990 the top ten held 69 percent. The FRC noted, however, that by global standards the manufacturing facilities of BC companies were small. It went on to echo Pearse's assertion: "The wood processing industry must be allowed to evolve into whatever structure is needed to maintain its ability to compete in world markets." The concern expressed was over control of timber harvesting rights.

Accompanying the consolidation, and in some respects driving it, was a relentless process of technical innovation, in the woods as well as the mills. Elsewhere in the world, industry desperately needed and wanted improved logging and sawmilling equipment; to remain competitive, BC loggers and mill owners were forced to buy the new machinery as it came on the market. Also, a whole new range of environmental requirements were coming into force, requiring that logging be done by means less disruptive to the forests, and that mills utilize all their timber rather than burning residues or pumping them out of pulp mills in the form of effluent.

> There's a real mix of trucks in the Cariboo. The Chilcotin pine suits itself to short-log in the sawmill economy because there's a lot of defect in that wood. The trees are short anyway. So that way you can get the crooks and the crooked parts out of them by taking it on a short-log basis. There's still a lot of long logging too. Another major change is in the technology of logging trucks. Around here you will see every conceivable type of logging truck. Five-axle, right up to the eight-axle B-trains. It's economics. If you can haul more wood on a truck, you pay more per hour for the truck but you can get the cost down. Williams Lake, I think, pioneered a lot of the technology in logging trucks. Seventeen years ago there were a few tri-axles here, six-axle trucks. Now there's very few five-axles and there's a helluva lot of six-, seven-, eight-axle trucks. That's because of short logging. You can pack three banks of short logs. The gooseneck type of trailer probably first started here. You can haul either long or short logs on it. Now a lot of that technology isn't useful when you get into the steeper country because you've got to get those trailers up onto the back of the truck to get them back to the bush. Slippery roads. In the east, around Horsefly and Likely. In the west you can use almost any configuration truck because you're on flatter ground.
> —John Mansell, BC director of forestry for Weldwood.

Throughout much of the last twenty years, a major influence on the course of the Interior industry has been insect attacks on forests and attempts to salvage the damaged timber. Between 1976 and 1986, the average annual volume of timber damaged by bugs was about 11 million cubic metres. Insect infestations increased through the mid-1980s, fell off until 1990, and then increased again. In 1995, according to the Ministry of Forests, about 2.5 million cubic metres were damaged.

Although most Interior species are susceptible to one type of insect or another, heavy infestations usually focus on a single species. In about 1975, for example, storms blew down large areas of trees in Bowron Lakes Provincial Park east of Prince George, creating ideal breeding conditions for spruce bark beetles. Park authorities rejected proposals to harvest the downed timber and, following several years of climatic conditions conducive to an increase in bark beetle populations, the infestation exploded. By the early 1980s a vast area had been affected and an immense fire hazard had developed in the dead and dying stands. An unprecedented salvage operation was mounted under the leadership of Northwood, which involved the logging and mill operations of most of the forest companies in Prince George and Quesnel. By 1987 more than 125,000 acres (50,000 ha) had been logged, in one huge clearcut. It was a controversial undertaking and it is still unclear whether it was the logging or some other cause that stopped the insect infestation. In any case, an enormous volume of timber was salvaged and the area was successfully reforested.

In the early 1980s an infestation of pine beetle began in the Chilcotin country west of Williams Lake. Another salvage operation was mounted, this time under more difficult circumstances. The timber was smaller and the infestation more scattered, and it was less economical to recover the wood. Several attempts failed, until Carrier Lumber moved large portable mills into the area. Just as Bill Kordyban had mounted the only successful salvage operation in the Peace reservoir, he was the only one to do so here. Unfortunately the Chilcotin venture ended in an acrimonious court case between Carrier and the provincial government over who was responsible for reforestation of the area. By the late 1990s, the Chilcotin infestation was still underway.

In the woods, there were hardly any loggers working on the ground by the 1990s. Hand fallers had been replaced by feller-bunchers, chokermen by grapple skidders. A variety of processors cut whole trees into delimbed logs at the road, then a knuckle boom loader put them on a truck, which took them to the mill yard. There they were unloaded with giant log stackers and fed into automated sawmills, from which they emerged as finished lumber, wrapped in bundles for shipment—without ever having been touched by a human hand.

> An interesting thing happened when we went from crawlers to skidders. The chokerman disappeared. The skidder operator, he'd jump off and set his own chokers. And that posed a problem a few years later. The chokerman was the training position. A young fellow could go out and set chokers and gradually learn how to run a Cat. And then develop onto other machinery. All of a sudden we had no chokers. So the entry level job into the woods was running the skidder. It became a bit of a training problem. You went onto a machine that had a lot of power and you were on some pretty dangerous ground. If you didn't know what you were doing, you could get into a jackpot. The Compensation Board told us we should wear seatbelts on skidders. That was a tough one to accept because the perception was you were better off to jump out of there.
> —*John Mansell*

The ultimate in mechanized operations was adopted at Slocan's Mackenzie operation, under manager Carl Baker, when it was still owned by Fletcher Challenge. Until 1994 this was a fairly conventional Interior logging operation using feller-bunchers and grapple skidders. Tree-length logs were trucked to Williston Lake, dumped in the water and towed to Mackenzie, where they were delimbed and bucked before being fed into the mills. Log towing ceased in the winter, when the lake froze. All logging was done in the winter; because of wet ground conditions, summer logging was not possible and not permitted. Keeping the large mill complex supplied with logs under these circumstances was a dicey business.

> In the early '60s I worked on the Coast and I can remember we used to say it took a million dollars to get off the beach. Now it's probably five million on the Coast, but that million had become a factor in the Interior. When skidders first came in to the Kootenays, you could go to the Timberjack dealer in Cranbrook with $800 and come out of there with a 404 Timberjack. That was all the down payment you needed. A good skidder now is $250,000 plus. You sure as hell won't get it for $800 down.
> —*John Mansell*

An operator feeds cants into a twin band saw at a West Fraser mill. WFT

In 1994, with the Forest Practices Code looming, Fletcher switched to an entirely new harvesting system, developed in Scandinavia and finding increased use outside BC. Ground-sensitive single-grip harvesters fell, delimb and buck logs at the stump. Computer-controlled processing heads on the harvesters automatically cut logs to whatever lengths the mills want at any given time. Spreading limbs before themselves as they work, the harvesters can operate almost year-round without damaging the forest floor. Equally light forwarders load and carry the logs to roadside, where they are trucked to the lake and loaded onto an enormous log barge, the Williston Transporter. This $12-million vessel, built by Finlay Navigation, is designed to plough through several feet of ice, and can haul logs to the mills year-round.

In total, this operation is one of the most technically sophisticated logging systems in the world—and one of the most expensive. The harvesters and forwarders cost half a million dollars each and require skilled workers to operate and maintain them. Judging by what is happening in other countries, this is the future of logging, but so far only a few of these machines are operating elsewhere in BC.

> The northern Interior I think was ahead of us in mechanized logging. We went up there to look at a few of those operations once we heard about it, the method of logging. We at Pope & Talbot got into it pretty quick but we weren't one of the pioneers in it as we were in the sawmill end of things. It took us a while because of the unionized aspect. Somebody else could have just said let's get that bloody equipment. And then, it was a substantial outlay in capital for equipment. Trying to convince sawmill people that you had to spend a few million dollars this year on some new logging equipment and you're changing the whole thing around—that didn't impress 'em at all. I fought that battle lots with the board, but we managed to pull it off. But we had a helluva job laying out our profit plan for the whole expenditure of mechanizing.
> —Bill Uphill

Most Interior mills have reached a comparable level of technological sophistication. As competitors elsewhere began to catch up with the Interior sawmill industry in the late 1960s, BC producers had to become more productive. They did so by cutting higher value products out of their logs in some cases, and generally by improving utilization. The 1980s was a decade of intense sawmill improvement through the use of scanners and computers that appraise the potential of a log almost instantly and, with minimal human intervention, cut it. Faster, thinner saw blades evolved and mills began to cut to one-thousandth of an inch tolerances. All of these innovations and others enabled mills to run at much higher speeds while cutting more accurately.

> We put the first scanners into our Williams Lake mill in 1984. Had trouble with them for a couple of years. It's really been in the last ten years that they've been refined and you've been able to put them in almost trouble-free. And that's made a big difference to everybody in terms of lumber recoveries and so forth. Also I think we've upgraded our product in the last ten years. We've been required to figure out ways to upgrade our product, find new niches in markets. Just off the top of my head, I guess that's been in the last ten years. And of course in British Columbia the focus of the last ten years has been to do anything you can to reduce your log costs. In the milling side that's really been sawing for recovery. Putting equipment in that gets more lumber out of each log that goes in, that's been the way our company has focussed. Watching those wood costs rise almost out of control, we've had to find ways to saw for recovery in our sawmills. So we've really converted our mills over the last eight or nine years to be very high recovery sawmills.
>
> The BC interior became, out of necessity, the low-cost producer of wood products in North America. And we've now become, as far as I can tell, the highest-cost or at least a very high-cost producer. That doesn't bode well for the future of the industry unless we can find ways to reduce those costs now and become competitive again. I shouldn't say we're not competitive, but our costs have increased dramatically, much more dramatically than the costs of any of the other operators that supply our markets.
> —Hank Ketcham

Changes in mill technology have completely altered the nature of sawmill work. No one handles logs or lumber now; it is all moved with machines. The setters, doggers and trim saw operators are all gone, replaced by computer-controlled equipment. Some mills do not even have sawyers, only "operators" watching the scanner screens and monitoring the computers' decisions.

The Dunkley mill is probably the most advanced of the Interior sawmills. It has 160 employees, yet less than one third are engaged in the actual sawing and planing. The others are largely support staff—engineers, electricians, computer technicians, machinists—who keep it finely tuned and running smoothly.

> Things changed completely from 1965 to 1985. That's the difference in night and day. In 1965, basically we [licensee foresters] were making every decision as long as it could be justified and was probably right. By 1985 you were just working to a set of guidelines or a set of rules that [the Forest Service] was putting out. The licences were renewed subject to you putting in an acceptable working plan. It wasn't really your working plan. You were creating a plan that they would find acceptable. Let's say you believed that you had a good system to use natural regeneration and a very quick initial spacing. If they didn't think it was free to grow with that initial spacing, well then you had to write the working plan to include their words to get the thing on a free-to-grow basis. I'm probably not saying that very logically, but it's an altogether different approach to who's managing, and whose decisions are working, and what is working the best in the forest. It was definitely far more interesting and exciting and, I think, probably more progressive in the '60s. But that's where we are now; you can't go back. Now it's even worse. It just gets worse all the time.
> —Fred Waldie

Back in the 1950s, a couple of guys with a power saw, a horse and a bush mill could log in the morning, saw lumber in the afternoon and go home that night with about 2,000 board feet of crudely cut lumber piled on their truck—a thousand feet each. Today, taking into account everyone employed at a high-end modern Interior sawmill, each worker produces about 4,400 board feet of finished lumber. About one-third of the labour goes into finishing—planing and drying. Not only is the lumber more valuable, but the residues produced in modern mills are also utilized, whereas those produced by the bush millers were burned or left to rot in the woods. There are proportionally fewer loggers working today than there were forty years ago, indicating that productivity has increased even more dramatically in the woods than in the mills. Since there are about the same number of people working in the industry today as forty years ago, the overall result of technological change has been an increase in the value extracted from the timber harvested.

During the past twenty years an entirely new sector of the forest industry has developed, much of it in the Interior. It is known as the value-added or remanufacturing sector. One of the first to get into this business was a young Dutch immigrant named John Brink who arrived in BC in 1965. He started piling lumber at a Watson Lake mill, and ended up as manager before moving to Prince George. He was inspired by Peter Pearse's report of 1976, which identified opportunities to create more valuable products out of the existing timber supply.

In 1975 Brink started Brink Forest Products. He bought economy-grade lumber from the high-speed sawmills in the city which were churning out commodity lumber for the US market. He ran this lumber through his mill, carefully cutting higher-grade boards out of the mass-produced material. Within a few years, he was employing 150 workers and a number of imitators had started operations.

Since then, about 400 remanufacturing firms employing 12,000 workers have opened for business in BC, some upgrading lumber produced by the primary sector, others manufacturing secondary products such as furniture and millwork. A portion of the timber available through the Small Business Forest Enterprise Program has been diverted to this sector, enabling remanufacturers to trade logs for lumber with the sawmills. Most of these secondary producers are members of the BC Council of Value-Added Wood Processors, which in 1998 elected Brink as its first president.

> In 1982, finally, I was awarded a small quota, 25,000 metres, which is not very big. It's about 5 million board feet. Depends how much you can cut out of a log. The setup they have now, the most modern mills, they only need about four metres to make a thousand board feet of lumber. I wasn't as efficient with saw kerfs and different things. I probably needed about five metres to make a thousand board feet. Then I moved to a sawmill site in town. I picked that year to move because prices really had dropped and I figured there's no money to be made in lumber, it's a good time to move and set up and be ready when the prices go up again. I started there in the fall of '82 sawing. In '83 I almost broke even. In '84 I lost $80,000. Well, I'm getting older and I worked hard to get where I got. At the end of your lifetime, to start losing money, that doesn't look very good, you know. I could afford it for one year, but not too many years. So then I looked at selling. And it didn't look like it would improve any at all. So then I approached PIR, that's West Fraser, and they bought my quota. At times, walking through the mill, something you built up and sweated for twenty-five years, you also got a lot of your emotional feelings in that, I was ready to call the deal off because I couldn't part with it. It was part of myself. I could hardly part with it. But it was good that I didn't call it off because lumber prices never improved much till the '90s. I got a logging contract from PIR and since '85 I've been logging.
> —John Veerbeek

Now, at the end of the twentieth century, the Interior forest industry has matured. By global standards it has reached a high level of technological sophistication and it has consolidated to the degree required to attain that sophistication.

Unfortunately the industry is also in the midst of the most severe crisis in its 140-year history. In a few short years it has gone from being North America's lowest-cost producer of wood products to one of the highest. The changing social values which have brought about increased government regulations, the increased stumpage charges due in part to US pressures, and the natural cost increase of having to log less accessible, less productive stands, have caused production costs to rise faster for BC Interior producers than for any other forest industries selling in global markets.

One measure of the severity of the situation is that in the mid-1990s, Scandinavian and western European competitors have begun selling forest products into both US and Japanese markets at prices BC mills cannot match. They are doing so under environmental standards as high as those required in BC. And all of the timber they are processing has been cultivated, while BC mills are still utilizing timber that grew without human effort or expense.

Thus, at the end of its first full century of existence, the Interior industry finds itself in a paradoxical situation. It is founded upon one of the most productive and valuable forest resources on the face of the earth. Taken as a whole, BC's Interior forests are unsurpassed in their diversity, their health and their economic worth. In most of its sectors, the industry is as well equipped as any in the world and the expertise of its workers is on a par with the best. But despite these advantages, the Interior forest industry is facing its future with uncertainty and doubt—not about its own potential and capabilities, but about the social, political and market forces that act upon it, and whether those forces will allow the industry to thrive as it has in the past.

Oral History Sources

David Ainsworth: Interviewed at Vancouver by the author, 1998.

R.E. Allen: Interviewed at Nelson by C.D. Orchard, 1959. UBC.

Walter Anchicoski: From an unpublished memoir. EDMS.

Jack Aye: BCARS.

Audrey Baird: Interviewed at his home by Enderby & District Museum staff, 1989.

Ike Barber: Interviewed at Vancouver by the author, 1998.

Bill Batten: Interviewed at Golden by the author, 1997.

Len Bawtree: Interviewed at Vernon by the author, 1997.

Peter Bentley: Interviewed at Vancouver by the author, 1997.

Orond Brashier: Interviewed at his farm at Parson by the author, 1997.

Jack Broderick: "Early Days of Logging in the Penticton and Princeton Area," unpublished manuscript, 1965. Penticton Museum.

Vic Brown: Interviewed at Cranbrook by the author, 1997.

G.O. Buchanan: From a speech to the Nelson University Club, May 9, 1908. Selkirk College Library.

Martin Caine: Interviewed at Prince George by C.D. Orchard, 1955. UBC.

Don Clutterham: Interviewed at Vancouver by the author, 1997.

Bob Cunningham: Interviewed at Crescent Valley by Peter Chapman, 1990 and 1993. KMA, *A Life in the Woods*, vol. I.

John Dahl: Interviewed by Robert Allington, 1995. *Business Logger*, January 1996.

Robert Dale: From a speech to the annual meeting of the Enderby & District Museum Society, November 4, 1992.

Allen H. DeWolfe: Interviewed at Victoria by C.D. Orchard, 1958.

John Dokkie: Interviewed by the author, 1997.

Al Dupilka: Interviewed at Quesnel by the author, 1997.

Sidney Eger: Interviewed at Vancouver by the author, 1997.

Cliff Falloon: Interviewed at Quesnel by the author, 1997.

Russell Fletcher: Interviewed by Peter Chapman, 1990. KMA, *A Life in the Woods*, vol. I.

Harry Gairns: Interviewed Mackenzie & District Museum staff, 1978.

R.J. Gallagher: Interviewed at Prince George by the author, 1997.

J. Miles Gibson: Interviewed at Fredericton by C.D. Orchard, 1960. UBC.

Ross Gorman: Interviewed at Westbank by the author, 1997.

Maitland Harrison: From Peter Chapman, *Where the Lardeau River Flows* (Sound Heritage Series, No. 32), BCARS.

L. Sawyer Hope: Interviewed at Victoria by C.D. Orchard, 1961. UBC.

Harold Jacobson: Interviewed at Williams Lake by the author, 1997.

Bruce Kerr: Interviewed at his farm by the author, 1997.

Hank Ketcham: Interviewed at Vancouver by the author, 1998.

George Lambert Interviewed at Nelson by Shawn Lamb, 1985; and by Peter Chapman, 1990. KMA, *A Life in the Woods*, vol. I.

Howard Lloyd: Interviewed at Prince George by the author, 1997.

Neil McLean: Interviewed at Kamloops by the author, 1997.

Don Mallard: Interviewed at Cranbrook by the author, 1997.

Bill Mann: Interviewed at Kamloops by the author, 1997.

John Mansell: Interviewed at Williams Lake by the author, 1997.

John Martens: Interviewed at Vanderhoof by the author, 1997.

Harry Meyers: Interviewed at Canal Flats by the author, 1997.

Len Mitten: Interviewed at Crescent Beach by the author, 1998.

Fred Mulholland: From a speech to the BC Loggers' Association at Vancouver, 1936. *Forestry Chronicle*, December 1954.

C.D. Orchard: From letters to Chief Forester P.Z. Caverhill, March 13, 1928 and February 20, 1930. BCARS.

Vivian Pederson: From her account in *Marks on the Forest Floor*, a story of Houston, BC, published by the Centennial '71 Committee.

Fritz Pfeiffer: Interviewed at Telkwa by the author, 1997.

Conrad Pinette: Interviewed at Vancouver by the author, 1998.

George Richards: Interviewed at Vancouver by the author, 1997.

Oscar Schmidt: Interviewed at Nelson by Peter Chapman, 1990. KMA, *A Life in the Woods*, Vol. I.

Bill Seafoot: Interviewed at Kaslo by the author, 1997.

Fred Stinson: Interviewed at Penticton by the author, 1997.

Archie Strimbold: Interviewed at his home west of Burns Lake by the author, 1997.

Jack Thompson: Interviewed by the author, 1997.

Al Thorlakson: Interviewed at Vernon by the author, 1997.

Bill Uphill: Interviewed at Midway by the author, 1997.

John Veerbeek: Interviewed at Smithers by the author, 1997.

Bill Waldie: Interviewed at Robson by the author, 1997; and by Peter Chapman, 1990. KMA, *A Life in the Woods*, vol. II.

Fred Waldie: Interviewed at Blind Bay by the author, 1997.

Dave Webster: Interviewed at Hazelton by the author, 1997.

Ray Williston: Interviewed at Gibsons by the author, 1998.

David Winton: *W's Back to Back*, published privately, 1971.

Joe Wrangler: Interviewed at Nelson by the author, 1997; and by Peter Chapman, 1993. KMA, *A Life in the Woods*, vol. II.

Tom Wright: Interviewed at Vancouver by the author, 1997.

Index

Abernathy Sawmills, 183
aboriginals, see Native British Columbians
Adams Lake, 42–43, 137, 161, 166; sawmilling, 125, 200, 201
Adams River, 42–43
Adams River Lumber, 37, 42–45, 81
Adolph Lumber, 41, 45, 63
Agur, Don, 183
Ainsworth family, 151, 155, 219–20
Ainsworth, David, 138–40, 140, 141, 166
Ainsworth, Tom, 138–40
Ainsworth Lumber, 138–40, 185, 219–20
Alaska Highway, 116
Alcan, 198
Alexander's, 154
Alexandra Forest Industries, 171, 179–80, 184–85
Aleza Lake experimental forest station, 166
Allen, R.E., 62, 65, 74
Altmonson Lumber, 110
Anchicoski, Walter, 142
Andersen, Ivan, 221
Anderson, Sivert "Bull River Slim," 104
Antler, 17
Apsey, Mike, 211–12
arch trucks, 133, 134–35, 137, 153, 159, 187, 188
Archibald Creek, 52
Argital, Harry, 226
Armstrong area: logging, 41; sawmilling, 27, 41, 53, 57, 118, 183, 220
Armstrong Sawmill, 41, 53, 57, 118, 183, 220
Arrow Lakes, 29, 47, 138
Arrow Lakes Lumber, 29, 74
Ashton Creek: sawmilling, 122
aspen, 216
Atlin: sawmilling, 34
auctions, 183
Avola, 28
Aye, Jack, 40, 41, 45

B&F Sawmills, 124, 152
Babine: sawmilling, 182
Babine Forest Products, 201, 224
Baillie-Grohman, William Adolph, 29, 32–33
Baird, Audrey, 111, 137
Baird Brothers Logging, 134, 142
Baker Lumber, 62
Balco Industries (Balco Forest Products), 125, 149–50, 218–19
Balestra & Schnyder, 183
Balfour-Guthrie, 201
Balison family, 218–19
Balison, Bert, 125, 148–50
Balison, Dave, 219
band saws, 26, 37, 42, 163, 195, 197–98
Barber, Ike, 198, 225–29
Barnes Lumber, 183
Barrett, Premier David, 207
Batten, Bill, 155, 169
Bawtree, Len, 115
Baylor, R.P., 17
Baynes Lake, 78; sawmilling, 63
BC Council of Value-Added Wood Processors, 235
BC Federation of Labour, 62
BC Forest Products, 171, 179–80, 199, 203, 215
BC Hydro, 185
BC Interior Sawmills, 116, 181
BC Loggers' Association, 130, 135
BC Railway, 181
BC Southern Railway, 33
BC Spruce Mills, 64, 69, 114, 116, 119, 182
BC Veneer Works, 82–83
Bear Creek: logging, 43–45
Bear Lake: sawmilling, 221
Beattie, Harold, 40
Beaver River, 23
Beaverdell, 122; sawmilling, 183
Beavermouth: sawmilling, 23, 24
Bednesti, 109

Bell, M.J., 50
Bell Pole, 50
Bellos Lake: logging, 154
Bene, John, 83, 154, 224
Benedict, R.E., 65
Bennett, Premier Bill, 209, 211, 229, 219, 221, 229
Bennett, Premier W.A.C., 145, 171
Bennett: sawmilling, 15, 34
Bentley family, 154
Bentley, Peter, 172, 174, 175, 178
Bentley, Poldi, 172, 174
Bergren, Hjalmar, 130
Bessette, Phos, 116, 181
Betsy (Countess of Dufferin) (locomotive), 24
Big Bend Lumber, 144
birch, 82
Black, Peter, 109
Blaeberry River, 27
Blue Mountain Sawmills, 116
Bluenose, 109
BNC Spruce Mills, 111
Bogue and Brown sawmill, 52
Bonners Ferry, 33
Boulder Creek, 52
Boundary Forest Products (Boundary Sawmills), 200
Boundary Sawmills, 122, 144, 198–201
Bowater Paper, 182
Bowman Lumber, 29, 45
Bowron Lakes area: logging, 231
box factories, 52–53, 81, 124
Boynton, J.R., 50
Brashier, Orond, 110, 131
Bridal Falls, 139
Bridges, Jonathan, 64
Bridges Lumber, 64
Brink Forest Products, 235
Broderick, Jack, 68
Brown, Roy, 159
Brown, Vic, 181–82
Brownmiller, Oliver, 124
Brownmiller Brothers, 124, 138, 152, 153
Bruhn, R.W., 121
Buchanan, G.O., 14, 32
Buchanan Lumber, 14, 29–32
Buck River Lumber, 109, 182
Bucknell, George, 111
Bulkley Valley, 104, 106, 107
Bulkley Valley Forest Industries, 224
Bulkley Valley Pulp & Timber, 109, 182
Bull River, 65–71, 78; sawmilling, 144
Burns, Don, 128
Burns Lake: sawmilling, 104, 182, 201, 224
Burns Lake Native Development Corporation, 201
bush mills, see portable sawmills
Butters, Horace, 99
Byspalko, Pete, 153

cable logging, 98–99, 135, 160
Caine, Martin, 83, 158, 178
Caine Lumber, 178
Calvert Planing Mills, 183
Camozzi, Vic, 121
Canadian Car Company (CanCar), 194, 195–97, 198
Canadian Cellulose, 138, 208
Canadian Forest Products (Canfor), 150, 154, 166, 172–78, 195, 201, 213, 214, 217, 219, 220–21, 228, 229, 230
Canadian Lumbermen's Association, 133
Canadian National Railway, 74, 83, 89
Canadian Northern Railway, 61
Canadian Pacific Railway, 19, 21, 27, 33, 40, 42, 45, 65–71, 74, 122
Canadian Pacific Timber, 45–46
Canadian Timber and Saw Mills, 45
Canadian Western Lumber, 24
Canal Flats, 32–33; sawmilling, 29, 71, 128
Canford: sawmilling, 50–51
Canoe: sawmilling, 121, 151
Canyon City Lumber, 68, 69

Cariboo Lumber Manufacturers Association, 131–32
Cariboo Pulp & Paper, 198, 224, 230
Caribou Lake: logging, 138
Carmi: sawmilling, 144
Carney (old-timer at Kaslo, early 1900s), 62
Carney Pole, 50, 121
Carrier Lumber, 184–86, 232
Cascade: sawmilling, 29
Castlegar area: logging, 46–47, 138; pulp milling, 144, 180, 208, 226, 229; sawmilling, 111–14, 138, 144, 180, 230
Caterpillars, see crawler tractors
Cattermole, Bob, 180, 228, 229
Caycuse: logging, 138
cedar, 219; see also western red cedar; yellow cedar
Cedar Creek: sawmilling, 163
Cedar Valley, 74
Cedar Valley Lumber, 29
Cedarvale, 104
Celanese Corporation, 142, 144
Celgar, 47, 112, 138, 144, 180, 208, 229, 230
Cetec, 195
Champion International, 224
Champion Papers, 182
Chapman, Brick, 119
Chase family, 42
Chase area: logging, 37, 42–45; sawmilling, 42–45, 50, 81, 121
Chase Lumber, 121
Chasm: sawmilling, 220
Cherryville, 54
Chetwynd: logging, 217, 227; sawmilling, 158, 197
Chetwynd Forest Industries, 218
Chip-N-Saw, 194, 195–98
circular saws, 16–27
Clark, Nick, 51–52, 53
Clear Lake: sawmilling, 221
Clearwater: logging, 52
Clearwater Timber Products, 144, 228
Clinton: sawmilling, 134; 138–39
Clinton Sawmills, 138–39, 220
Clive, 33
close utilization, 194–95, 202
Clucultz Lake area, 154; logging, 158
Clutterham, Don, 201, 203
Columbia and Kootenay Railway, 29
Columbia and Western Railway, 33
Columbia Cellulose, 142, 143
Columbia Lake, 32–33
Columbia River, 23, 28–29, 32–33, 230
Columbia River Lumber, 24, 26, 27, 34–37, 41, 69, 74, 77, 78, 110
Columbia River Timber, 144
Commercial Lumber, 201
Commission of Conservation (1918), 78
Commission on Resources and the Environment, 215
Comox Logging & Railway, 17
Connelly, Mark, 109
conservationism, 37–40, 58–59, 78
Consolidated-Bathurst, 182
Cooke Creek, 134
Cooke Lumber, 183
Cooper-Widman, 182
cottonwood, 46, 82
Cottonwood: logging, 153
Council of Forest Industries, 132, 211, 212
Countess of Dufferin (Betsy) (locomotive), 24, 71
Craigellachie, 24
Cranberry River, 230
Cranbrook area: logging, 40, 64–71; pulp milling, 181–82; sawmilling, 33, 46, 62, 63–66, 74, 76, 100, 101, 128, 163, 181–82, 224
Cranbrook Lumber, 33
Cranbrook Sash and Door, 33, 46, 116
Cranbrook Sawmills, 128, 144
crawler tractors, 91–98, 96–98, 119, 122, 128, 133–34, 139, 149, 150, 152, 158, 159, 161, 162, 185, 186, 188, 203, 232

Crescent Valley, 52, 55; sawmilling, 50
Crestbrook Forest Industries (Crestbrook Timber), 128, 129, 181–82, 189, 208, 224–25
Creston area: logging, 68; sawmilling, 46, 68, 128
Crooked River: logging, 162
Crossing Lumber, 183
Crows Nest Industries (Crow's Nest Pass Coal; Crows Nest Pass Lumber), 33, 62, 63, 125, 224–25
Crown Zellerbach, 118, 125, 172, 181, 183, 208, 220
Cunningham, Bob, 134
Curly (locomotive), 22

Dahl, John, 181
Daishowa, 182, 218, 221, 224
Dale, Mel, 206
Dale, Robert, 53, 54, 57
Dalskog, Ernie, 130
Davey Lake: logging, 162
Davison, Merv, 174
Dawson Creek: sawmilling, 116, 150, 218
Deadwood: sawmilling, 29
Decker Lake: sawmilling, 104
Deejay Trucking, 212
DeGrace, Larry, 166, 168, 169m, 171, 172, 174
Demaid, Bill, 143
Depression, 109–14, 116
DeWolfe, Allen H., 98
Dixon, Al, 183
Dokkie, John, 105
Donald: logging, 24; sawmilling, 27, 118–19, 201
Donohue company, 229
Douglas fir, 12, 37, 43, 46, 50, 53, 66, 77, 122, 130, 147, 163, 183, 206
Downie Street Sawmill, 121
Drew Sawmills, 183
Driscoll Creek: logging, 169
Drott, Erv, 189
Duke, Pat, 125
Dumont family, 202, 218
Dunkley Lumber, 159–60, 234
Dupilka, Al, 122

Eagle Lake: sawmilling, 172
Eagle Lake Sawmills (Eagle Lake Spruce Mills), 83–85, 91–95, 98, 101, 116, 130
Eagle Pass, 23–24
East Kootenay Lumber, 63, 74
East Line, 83–151, 130, 131, 147, 155, 172, 221
Eastern BC Lumber, 74
edgers, 26, 42, 46
Edgewood: sawmilling, 46
Edgewood Lumber, 112
Eger, Sidney, 153, 183
Elco, 46
Elk Lumber, 24, 29
Elk River Lumber, 62
Elko, 78; logging, 24; sawmilling, 24, 62, 125, 225
Ellett Lumber, 183
Ellis Creek: sawmilling, 41
Emory, 21, 22
Endako: sawmilling, 103, 104
Enderby area: logging, 52–54, 57, 58, 122, 134, 137, 142; sawmilling, 28, 34, 52–58, 81, 118, 121–22, 141, 148, 183, 206, 220
Enterprise, 14, 17
environmental issues, 209, 213–15, 231, 233
Erickson, Axel, 225
Ernst, John, 122, 195, 218, 220
Ernst Lumber, 195, 220
Eurocan Pulp & Paper, 198, 199, 201, 218
European exploration, 12
Evans Forest Products, 119, 201, 220
Exco Industries, 220
Exeter, 219
exports of forest products, 212–13, 215, 216, 217, 218

Fairmont: sawmilling, 128
Falkland: sawmilling, 183, 220
Falloon, Cliff, 124, 138, 152, 186, 188, 202
Falloon Brothers Sawmill, 124, 152, 153
Fanshaw (early owner of Sinclair Mills), 84
Farstad, Alf, 128, 129, 181–82
Federated Co-operatives, 121, 125, 151, 183
feller-bunchers, 189, 203, 207, 216, 217
Fensom, George, 19
Fernie area: logging, 29; sawmilling, 29, 33, 41, 45, 62–63, 65, 74
Fernie Lumber, 74
Fibreco Export, 228, 229
fin booms, 86, 100
Finlay Forest Industries, 180, 181, 229
Finlay Forks: sawmilling, 180, 184
Fisher Planing Mills, 183
Fitz Sawmill, 200
Flagstone, 78; sawmilling, 63
Fletcher, Russell, 51, 52
Fletcher Challenge, 220, 229, 233
Floyd, Doug, 198
flumes, 18, 43–45, 47, 52, 69–70, 111, 114, 115, 116, 119
Flynn, Don, 224
Foote, Charlie, 103
Foote, Howard, 103
Forbes, Jack, 103
fore-and-aft roads, 94–95
foreign investment, 171, 180, 181–82, 208, 218, 221, 224, 225
Forest Act (1912), 59; (1978), 165, 211, 213
forest fires, 58, 59, 77–78
Forest Licences, 211
Forest Management Licences, 141–44, 179
Forest Practices Code, 215–17, 233
Forest Renewal BC, 217
Forest Resources Commission, 215
Forest Service, 17, 59, 74, 78, 82, 83, 97, 105, 109, 116, 121, 122, 141, 142, 158, 163–65, 174, 184, 186, 193, 203, 207, 209, 211–12, 213–14, 234
Fort Fraser: sawmilling, 86, 103
Fort Nelson area: logging, 203; pulp milling, 228; sawmilling, 116, 180, 228, 229
Fort Nelson Forest Industries, 228
Fort Nelson Indian Band, 228
Fort St. James area: logging, 189, 217; sawmilling, 230
Fort St. John area: logging, 188, 213; sawmilling, 116, 150, 197
Fort St. John Lumber, 116, 158
Fort Steele area: logging, 64, 130; sawmilling, 130
Fortune, William, 18, 19
Four Mile Creek, 41
Fowler, Halley, 159
François Lake: logging, 109; sawmilling, 106
Fraser Lake: sawmilling, 109, 198, 218
Friend, Albert, 139
Frost (mill owner at Willow River, c. 1915), 83

Gairns, Harry, 184, 185
Galbraith Creek, 65
Galena Bay: sawmilling, 74
Gallagher, R.J., 131, 132, 133
Galloway: sawmilling, 125
Galloway Lumber, 125, 144
gang saws, 26, 140, 195, 198
Ganzeveld, Engel, 122–23
Ganzeveld, Rex, 122–23
Garland, Curt, 228
Genelle family, 37
Genelle, Addie, 29
Genelle, Jack, 29
Genelle, Joseph, 28, 29
Genelle, Peter, 28, 29
Genelle, Sadie, 29
geology of BC, 11–12
Gerrard, 45–46

INDEX

Getz, Roger, 207
Gibbons Lumber, 74
Gibson, J. Miles, 83, 109
Ginter, Ben, 180
Giscome: sawmilling, 83, 91–95, 98
gold rush, 12–14, 34–35
Golden area: logging, 22, 71; sawmilling, 22, 24–27, 33, 74, 77, 78, 81, 124, 144, 201
Golden Lumber, 124, 182–83
Goodlett, Lorne, 228
Gopp, Henry, 112
Gorman, John, 125
Gorman, Ross, 125
Gorman Brothers Lumber, 125
government, *see* Forest Service
Grand Forks Sawmills, 200
Grand Trunk Pacific Railway, 51, 61–62, 83, 103; *see also* Canadian National Railway
Great Northern Railway, 33
Green, A.H., 52
Greenwood: sawmilling, 183
Grindrod: pole cutting, 50; sawmilling, 132
Grindrod Lumber, 132
grist mills, 19, 27
Grohman Creek: logging, 160

Hagman, Harry, 107, 109, 182
Halcro, Stan, 139
Hales-Ross Planing Mill, 85
Halfway: sawmilling, 141
Hallaln, Bruce, 159
Hanbury, Wilf, 183
Hanbury, 78; sawmilling, 63
Hanson Lumber and Timber, 51, 104, 111
Hanson, Olof "Tie," 74, 103–4, 107, 109
Harcourt, Premier Mike, 215
Harper, Jerome, 17
Harris, Ormie, 121
Harrison, Maitland, 46
Hawes, Charlie, 115
Hazelton area: logging, 65, 212; sawmilling, 51, 104
head saws, 26, 77
Heffley Lake area: sawmilling, 125, 148, 218
Helcro Forest Products, 221
Helen, 42, 43
hemlock, 12, 37, 46, 104, 183
Hidden Lake: logging, 142
Higginson, Thomas, 27
high-lead logging, 135, 160
Hill Brothers, 29
Hixon area: sawmilling, 159–60
Holding, Art, 125, 159, 181, 201
Holding Lumber, 125, 137, 148, 159, 166, 200, 201
Honshu, 181
Hood Lumber, 74
Hoover Sawmill, 202
horse logging, 24–26, 34, 37, 40, 41, 43, 47, 50, 51, 53, 65–69, 74, 75, 84, 86, 87, 89, 90, 91, 100, 107, 111, 119, 128, 130, 131, 142, 150, 159, 161, 186, 203
Horse Thief Creek, 69
Horsefly: logging, 231; sawmilling, 161–63
House, Albert, 19
Houston: pulp milling, 182; sawmilling, 107, 109, 145, 182, 224
Houston Forest Products, 224
Howe, C.D., 115
Howser, 46
Hudson's Bay Company, 12, 17
Hunter, Bill, 62
Hutton: logging, 86; sawmilling, 83

Illicillewat River, 23
Industrial Forestry Service, 168, 171, 184
Industrial Workers of the World (Wobblies), 62, 129
insect damage, 231–32
Intercontinental Pulp, 178
Interior Lumber Manufacturers Association, 131–32
International Woodworkers of America (IWA), 129–31
Isle Pierre: sawmilling, 228

Jacobs (flume builder at Boulder Creek, early 1900s), 52
Jacobson, Harold, 161–63
Jacobson, O.J., 161–63
Jacobson Brothers Forest Products, 161–63, 220
Jaffray, 78; logging, 74, 75; sawmilling, 63, 74
Jamieson, James, 18
Jewel, Gordon, 107
Jewel Lumber, 63
John, Bill, 197
Johnson, Ches, 226
Johnson, Ken, 124
Johnson, Rudy, 168
Johnston, Bill, 12
Johnston, Eli, 12
Jones (superintendent at Potlatch Lumber), 98
Jones, Bill, 143
Jones Creek: sawmilling, 139
Jorgenson and Wells, 139
Jostad, Oscar, 125
Jura: logging, 68

Kalesnikoff family, 129
Kalesnikoff Lumber, 129
Kamloops area: logging, 17; pulp milling, 181; sawmilling, 18–19, 22, 27–28, 125, 181, 183, 201, 218
Kamloops Lake, 42
Kamloops Pulp & Paper, 124, 125, 181
Kamloops Sawmill, 28
Kaslo: logging, 62; sawmilling, 14, 29–32
Kelowna area: logging, 159, 172; sawmilling, 34, 81, 124–25, 144, 172, 183, 195, 220
Kelowna Lumber, 81
Kelowna Sawmill, 34
Kenney, Edward, 142
Keremeos: sawmilling, 183
Keremeos Lumber, 183
Kerr, Bruce, 107, 128, 134, 164
Kerr, Leslie, 153, 154–55
Kerry Lake: sawmilling, 195
Ketcham family, 122, 150, 155
Ketcham, Bill, 155–58
Ketcham, Hank, 234
Ketcham, Pete, 155–58
Ketcham, Sam, 155–58, 195–97, 198, 218
Kettle Valley: sawmilling, 122, 144
Kettle Valley Railway, 50–51
Kicking Horse Forest Products, 124, 181, 182–83, 201
Kicking Horse River, 22
Killy, George, 221
Killy, Ivor, 174
King, A.S., Logging, 153, 154
Kingfisher Sawmills, 183
Kitchener, 78; sawmilling, 64
Kitimat area: logging, 198; pulp milling, 198
Koch, W.C.E., 50
Kootenay and Arrowhead Railway, 45
Kootenay Forest Products, 138, 160, 203, 208, 229
Kootenay Lake, 29–32
Kootenay River, 32–33
Kordyban, Bill, 184–86, 232
Kualt: sawmilling, 24, 28, 29

Lady Dufferin, 19
Lafferty Creek, 184
Lafon (Forest Service organizer, early 1900s), 65
Laggan, 22
Lakeland Mills, 221
Lamb, C.R., 28
Lamb-Watson Lumber, 28, 29, 41
Lambert, George, 47
Lambert Lumber, 47
laminated veneer board plants, 201, 220, 230
Land, Fleming and Company, 15, 16
larch, 46, 53, 82, 130
Lardeau River, 45, 46
lath mills, 26, 46, 52–53, 71
Lavington: sawmilling, 218, 220
Lavington Planer Mill, 201
Leary, Sid, 62
Leask, Tom, 46
Leir, Hugh, 41, 68, 111, 116, 183
Leith, Chuck, 188
Lequime, Bernard, 34
LeTourneau, Bob, 97
Liebscher, P.F., 163
Liersch, John, 172, 174
Lignum Ltd., 140, 155, 183
Likely: logging, 231; sawmilling, 174
Lillooet area: sawmilling, 201, 220
Lindsley Brothers Pole, 50
Little Slocan River, 50, 52
Lloyd, Howard, 154, 158, 162, 202, 224
Lloyd-Jones, David, 34, 81
loaders (jammers), 26, 71, 74–75, 76, 172, 203, 214
lodgepole pine, 12, 103, 104, 139–40, 141, 155, 163, 193, 201, 219
logging: Armstrong area, 41; Bear Creek, 43–45; Bellos Lake, 154; Bowron Lakes area, 231; Canal Flats, 32–33; Caribou Lake, 138; Castlegar area, 46–47, 138; Caycuse, 138; Chase area, 37, 42–45; Chetwynd, 217, 227; Clearwater, 52; Cluculz, 158; contract, 206–7; Cottonwood, 153; and CPR, 21–23; Cranbrook area, 40, 64–71; Creston area, 68; Crooked River, 162; Davey Lake, 162; Driscoll Creek, 169; Elko, 24; Enderby area, 52–54, 57, 58, 122, 134, 137, 142; Fernie area, 29; and fisheries, 43; Fort Nelson area, 203; Fort St. James area, 189, 217; Fort St. John area, 188, 213; Fort Steele area, 64, 130; François Lake, 109; Golden area, 22–23, 71; Grohman Creek, 160; Hazelton area, 65, 212; Hidden Lake, 142; Horsefly, 231; Hutton, 86; Jaffray, 74, 75; Jura, 68; Kamloops area, 17, 28; Kaslo, 62; Kelowna area, 159, 172; Kitimat area, 198; Likely, 231; Lumby, 125; Mabel Lake, 12, 40, 56, 137; Mackenzie area, 181, 215, 232; McBride, 155; Nelson area, 52, 78, 82; Noisy Creek, 137; Okanagan Lake, 172, 173; Parson, 110; Peace River dam, 181, 184–85; Penny, 155, 169; Penticton area, 41–42; Prince George area, 52, 82, 86, 116, 160, 232; Quesnel area, 17, 160, 202, 216, 232; Revelstoke area, 45; Salmon Arm, 142; Scandinavian system, 213, 233; Shelley, 155; Shelter Bay, 230; Shuswap area, 17, 42, 50, 53–54; Silverton, 62; Slim Creek, 155; Slocan area, 189, 214; Smithers area, 186; Sovereign Creek, 188, Summit Lake, 62, 162; Taghum, 47; Takla, 214; Telkwa, 107; Terrace area, 104; Torpy Creek, 155; Tum Tum Lake, 42; Usk, 104; Vernon area, 115; Williams Lake area, 139, 207, 231, 232; Williston Lake area, 199, 203, 232; Yale area, 21; *see also* horse logging; pole cutting; railway logging; river driving; tie cutting; timber supply
Lone Butte: sawmilling, 201
Long, Ken, 116, 181, 219
Long Lake Lumber, 183
Longworth, 116
Louis Creek: sawmilling, 219
Lower Arrow Lake, 46
Ludwig, Mike, 137
Lumberton (Wattsburg), 64, 69, 78, 116, 119; sawmilling, 111
Lumby: logging, 125; sawmilling, 123–24, 183
Lumby Timber, 123–24
Lunde, Bob, 228

Mabel Lake: logging, 12, 40, 56, 137; sawmilling, 34, 123
McAmmond's sawmill, 134
McBride, Premier Richard, 40, 58, 61–62
McBride: logging, 155
McDiarmid, Harvey, 124
McDiarmid, Ian, 124
MacDonald Lumber, 125
McDougall, Jack, 105
McGoldrick, J.P., 45
McGoran, Andrew, 50–51
McGuinness, George, 130
McIntosh, James, 18, 19, 27–28
Mackenzie area: logging, 181, 215, 232; pulp milling, 180; sawmilling, 180, 181, 184–85, 199, 203, 229
Mackie (foreman at Cranbrook mill, early 1900s), 40
McKinnon, Fin, 174
McLean, Neil, 86, 90, 96, 98, 100, 101
McLean, Sinclair, 86–91, 101, 103, 124
McLean Lumber, 86–91, 96
McMann, Bill, 171
MacMillan, H.R., 59, 65, 115
MacMillan Bloedel, 172, 198
McMynn, Gordon, 200
McPhee, Don, 84, 130
McQueen, Isaac
McVicker, Fred, 82
Mahood, Ian, 174–75, 225–26
Malakwa: sawmilling, 124
Malkin, Robert E., Lumber, 153
Mallard, Don, 130
Malpass, Gerald, 122
Malpass, Tom, 111, 122
Malpass Lumber & Poles, 123
Manistee, 78; sawmilling, 62
Mann, Bill, 94, 96
Mann, Harold, 91
Mansell, John, 231, 232
Mara, John, 28
Mara Lake, 53
Marten, 17
Martens family, 208, 229
Martens, Dave, 116
Martens, John, 152, 161, 187
match block mills, 51, 82
Mead company, 178
Meadow Lake: sawmilling, 139
Meeker, Henry, 51, 98, 109–11
Melrose, George, 82
Meltzer, Abe, 225–29
Merritt area: sawmilling, 19, 50–51, 74, 98, 109–11, 116, 124, 219
Merritt Diamond Mills, 124
Meyers, Harry, 189, 209
Midge, 33
Midway: sawmilling, 198–200, 201
Milnar, Vinnie, 172, 174
Mission: sawmilling, 111
Mitsubishi, 181
Mitten, Len, 194, 195, 197, 198
Moberley Lake: sawmilling, 105
Moffat, Harold, 172
Monarch Lumber, 42
Monashee Lumber, 183
Monte Lake: sawmilling, 144, 183, 220
Moore, Gordon, 116, 197
Mount Hermit, 22
Mountain Creek bridge, 23
Mountain Lumber Manufacturer's Association, 77
Mountain Lumbermen Association, 131
Moyie River, 64, 111
Mulholland, Fred, 17, 78
Mundy Lumber, 74

Nakusp: sawmilling, 28, 29, 50, 144
Nason, Ithiel, 16, 17
National Forest Products, 173–74
Native British Columbians, 33, 201, 228
Nechacco, 52
Nechako River, 52
Nelson, Charlie, 125
Nelson area: logging, 52, 78, 82; sawmilling, 47, 208, 229
Netherlands Overseas Mills, 201, 221
New Denver Lumber, 29
Newport Creek: sawmilling, 202, 218
Nicholson, Allan, 46, 116
Nicholson, 24–26
Nicola Valley, 187; sawmilling, 19, 109
Nicola Valley Pine Mills (Nicola Valley Lumber Company, Nicola Valley Sawmills), 50–51, 74, 75, 98, 109–11, 116, 219
Noble, Curly, 123
Noisy Creek: logging, 137
Noranda, 173–78, 183
North Central Plywood, 224
North Star Lumber, 74
Northern Development Company, 51–52
Northern Interior Lumber Association, 131, 132
Northern Lumber, 83
Northern Pacific Railway, 33
Northwood Pulp & Timber, 109, 151, 178, 182, 183, 224, 231–32
Notch Hill: sawmilling, 24
Nova Lumber, 219
Novak family, 160
Novak, Tony, 160

Ocean Falls: pulp milling, 208
Okanagan Falls: sawmilling, 183
Okanagan Lake, 41, 172, 173
Okanagan Landing: sawmilling, 34
Okanagan River, 41
Olinger Lumber (Sawmills), 144, 200
Oliver Sawmills, 144
Olsen, Pete, 105
Onderdonk, Andrew, 21–22
One Big Union, 74
100 Mile House area: sawmilling, 141, 219
Orchard, C.D., 105, 116, 141
oriented strand board plants, 216, 219, 220, 229
Osoyoos: sawmilling, 183
Osoyoos Sawmills, 183
Ottertail Creek bridge, 23
Overseas Spruce Sales, 159

Pacific Great Eastern Railway, 61–62, 144–45, 159, 219
Pacific Pine, 225
Pacific Western, 122
Palliser, 22, 23, 27
Parker, Dave, 215
Parkin, Jim, 118–19
Parson: logging, 110; sawmilling, 110, 131
Passmore Lumber, 138, 144, 225
Patrick Lumber, 52, 55
Peace River dam, 171, 180, 181, 184–85, 199, 232
Peace Salvage Association, 184
Peachland Sawmill & Box, 124
Pearse, Peter, 209, 212, 213, 231
Pederson, Vivian, 145
Penny: logging, 155, 169; sawmilling, 83
Penny Spruce Mills (Sawmills), 83, 169
Penticton area, 68; sawmilling, 41–42, 183
Penticton Sawmills, 41, 183
Pfeiffer, Fritz, 128, 137, 189
pine, 45, 50, 53, 104, 133, 168, 231; *see also* lodgepole pine; ponderosa pine; white pine
Pinette, Conrad, 140, 141, 168
Pinette, Gabe, 140, 155, 168
Pinette and Therrien Planer Mills, 140, 155
pit sawing (whipsawing), 15
plank roads, 94–95
Plateau Mills, 116, 208, 229
plywood plants, 82–83, 139, 154, 201, 216, 220, 224, 228
pole cutting, 50, 51, 81–82, 104, 109, 111, 121, 122, 124, 130, 206
pole roads, 87–89
Pomeroy, Dan, 116
ponderosa pine, 12, 37, 46, 66, 77, 142
Ponderosa Pine Lumber, 144, 183, 220
Pope & Talbot, 122, 200, 233
portable sawmills (bush mills), 81, 116, 122, 125, 129, 140, 145, 147, 150–55, 158–59, 163–65, 166, 171, 172, 174, 180, 186, 193, 201, 232
Postill, Alfred, 34
Potlatch Lumber, 98
Powell, W.W., 51, 82
Powell River Company, 172, 198
power saws, 136–38, 139, 142, 186–87, 203
Price, Ron, 226, 227
Prince George area, 61–62; logging, 52, 82, 86, 116, 160, 232; pulp milling, 169, 172–78, 174, 178, 182, 230; sawmilling, 52, 53, 83, 84, 85, 86–95, 98, 131, 132, 151, 154, 158–59, 166, 168, 172–78, 180, 184, 201, 207, 221, 224, 235
Prince George Economic Development Commission, 172
Prince George Pulp & Paper, 178, 230
Prince George Wood Preserving, 224
Prince Rupert: pulp milling, 142, 208, 229; sawmilling, 104
Princeton: sawmilling, 19, 183
Pritchett, Harold, 130
pulp milling: Castlegar area, 144, 180, 208, 226, 229; Cranbrook area, 181–82; Fort Nelson area, 228; Houston, 182; Kamloops area, 181; Kitimat area, 198; Mackenzie area, 180; Ocean Falls, 208; Prince George area, 169, 172–78, 174, 178, 182, 230; Prince Rupert, 142, 208, 229; Quesnel area, 182, 198, 224; Taylor, 229
Pulpwood Harvesting Agreement, 175, 178, 179, 181, 182
Push, Pull & Jerk Railway, 47

Quadra Logging, 124, 152, 202
Quesnel area: logging, 17, 16, 202, 216, 2320; pulp milling, 182, 198, 224; sawmilling, 17, 51–52, 122, 124, 139, 140, 144, 150, 151, 154, 195, 198, 201–2, 207, 216, 218, 219, 220, 224, 225, 227
quotas, *see* timber quotas

Raboch, Gerald, 183
Raboch, Jerry, 121–22
Radium: sawmilling, 128, 201, 227
railway logging, 37, 71–74, 84, 119; Fernie area, 63; Jaffray, 75; Nicholson, 24–26; Wycliffe area, 64–65; Yale area, 22
railways, *see* British Columbia Southern Railway; Canadian National Railway; Canadian Northern Railway; Canadian Pacific Railway; Columbia and Kootenay Railway; Columbia and Western Railway; Grand Trunk Pacific Railway; Great Northern Railway; Kettle Valley Railway; Kootenay and Arrowhead Railway; Northern Pacific Railway; Pacific Great Eastern Railway; Push, Pull & Jerk Railway; Shuswap and Okanagan Railway; St. Mary's and Cherry Creek Railway
Rayonier, 208
Red Mountain Lumber, 83
Reed Paper, 174, 178
reforestation, *see* timber supply
Reid, James, 17
Repap, 219
Revelstoke area: logging, 45; sawmilling, 45, 74, 121, 144, 153
Revelstoke Building Materials, 45
Revelstoke Sawmills (Lumber), 128, 201, 218, 227
Richards, George, 216
Richfield, 17; sawmilling, 16
river driving, 28, 54–58, 70, 89, 109, 155
Riverside Forest Products, 115, 122, 123, 163, 183–84, 220, 228
Riviere, Jack, 159
Robbins, Ralph, 174
Robson: sawmilling, 29
Rock Creek: sawmilling, 29
Rogers, A.R. Lumber, 52–58, 118, 128
Rogers Pass, 22
Ross, Hales, 74
Ross, James, 23
Ross-Saskatoon Lumber, 63, 74
Rossland: sawmilling, 29
Royal Bank, 229
Royal City Planing Mills, 22
Royal Commissions, 215; 1909 (Fulton), 58–59; 1945 (Sloan), 141; 1975 (Pearse), 209, 211, 212, 231
Runyan, Ernie, 194, 195, 197
Rustad, Mel, 132
Rustad Brothers Planing Mill, 132, 224
Ruttan, John, 103
Ryans-Shields Sawmill, 28

S & O Sawmills, 201
St. Laurent, Lyle, 201
St. Mary's and Cherry Creek Railway, 64
Salmo: sawmilling, 51
Salmon Arm: logging, 142; sawmilling, 118, 122, 183
Salmon Arm Lumber, 183
salvage operations, 184–86, 231–32
San Jose Logging, 207
Saskatchewan, residents in BC forest industry, 152
Sauder family, 201
Savona: sawmilling, 42, 220
Savona Timber, 201
sawmilling: 100 Mile House area, 141, 219; Adams Lake, 125, 200, 201; Armstrong area, 27, 41, 118, 183, 220; Ashton Creek, 122; Atlin, 34; Avola, 28; Babine, 182; Baynes Lake, 63; Bear Lake, 221; Beaverdell, 183; Beavermouth, 23, 24; Bennett, 15, 34; Bull River, 71, 144; Burns Lake, 104, 182, 201, 224; Canal Flats, 29, 71, 128; Canford, 50–51; Canoe, 121, 151; Carmi, 144; Cascade, 29; Castlegar area, 46, 111–14, 138, 144, 180, 230; Cedar Creek, 163; Chase area, 42–45, 50, 81, 121; Chasm, 220; Chetwynd, 158, 197; Clear Lake, 221; Clinton, 134, 138–39; Cranbrook area, 33, 46, 62, 63, 74, 76, 100, 128, 163, 181–82, 224; Crescent Valley, 50; Creston area, 46, 68, 128; Dawson Creek, 116, 150, 218; Deadwood, 29; Decker Lake, 104; Donald, 27, 118–19, 201; Eagle Lake, 172; Edgewood, 46; Elko, 24, 62, 125, 225; Ellis Creek, 41; Emory, 21, 22; Endako, 103, 104; Enderby area, 28, 34, 52–58, 81, 118, 121–22, 141, 148, 183, 206, 220; Exeter, 219; Fairmont, 128; Falkland, 183, 220; Fernie area, 29, 33, 41, 45, 62–63, 65, 74; Finlay Forks, 180, 184; Flagstone, 63; Fort Fraser, 86, 103; Fort Nelson area, 116, 180, 228, 229; Fort St. James area, 230; Fort St. John area, 116, 150, 197; Fort Steele area, 130; François Lake, 106; Fraser Lake, 109, 198, 218; Galena Bay, 74; Galloway, 125; Giscome, 83, 91–95, 98; Golden area, 22, 24–27, 33, 74, 77, 78, 81, 124, 144, 201; Greenwood, 183; Halfway, 141; Hanbury, 63; Hazelton area, 51, 104; Heffley Lake area, 125, 148, 218; Hixon, 159–60; Horsefly, 161–63; Houston, 107, 109, 145, 182, 224; Hutton, 83; Isle Pierre, 228; Jaffray, 63, 74; Jones Creek, 139; Kamloops area, 18–19, 22, 27–28, 125, 181, 183, 201, 218; Kaslo, 14, 29–32; Kelowna area, 34, 81, 124–25, 144, 172, 183, 195, 220; Keremeos, 183; Kerry Lake, 195; Kettle Valley, 122, 144; Kitchener, 64; Kualt, 24, 28, 29; Lavington, 218, 220; Likely, 174; Lillooet, 201, 220; Lone Butte, 201; Louis Creek, 219; Lumberton, 111; Lumby, 123–24, 183; Mabel Lake, 34, 123; Mackenzie area, 180, 181, 184–85, 199, 203, 229; Malakwa, 124; Manistee, 62; Meadow Lake, 139; Merritt area, 19, 50–51, 74, 98, 109–11, 116, 124, 219; Midway, 198–200, 201; Mission, 111; Moberley Lake, 105; Monte Lake, 144, 183, 220; Nakusp, 28, 29, 50, 144; Nelson area, 47, 208, 229; Newport Creek, 202, 218; Nicola Valley, 109; Notch Hill, 24; Okanagan Falls, 183; Okanagan Landing, 34; Oliver, 144; Osoyoos, 183; Palliser, 22, 23, 27; Parson, 110, 131; Penny, 83; Penticton area, 41–42, 183; Prince George area, 52, 53, 83, 84, 85, 86–95, 95, 98, 131, 132, 151, 154, 158–59, 166, 168, 172–78, 180, 184, 201, 207, 221, 224, 235; Prince Rupert, 104; Princeton, 19, 183; Quesnel area, 17, 51–52, 122, 124, 139, 140, 144, 150, 151, 154, 195, 198, 201–2, 207, 216, 218, 219, 220, 224, 225, 227; Radium, 128, 201, 227; Revelstoke area, 45, 74, 121, 144, 153; Richfield, 16; Robson, 29; Rock Creek, 29; Rogers Pass, 22; Rossland, 29; Salmo, 51; Salmon Arm, 118, 122, 183; Savona, 42, 220; Shelley, 86–91, 101, 172; Sicamous, 42; Silverton, 50; Slocan area, 29, 50, 144, 221; Smithers area, 104, 128, 186, 198, 218; Sproule Creek, 47; Summerland, 183; Summit Lake, 50; Taft, 74; Taylor, 221; Telkwa, 128, 186; Three Valley Gap, 74; Tranquille, 18; Vanderhoof, 116, 208, 229; Vernon area, 122–23, 183, 220; Waldo, 62, 74; Wardner, 63; Watson Lake, 235; Westbridge, 201; Wildhorse, 17; Williams Lake area, 140, 141, 155, 158, 162–63, 166, 168, 197, 198, 218, 220, 234; Willow River, 83; Wineau, 29; Winfield, 34; Wycliffe, 74; Wynndel, 64, 111; Yahk, 106; Yale area, 15, 22, 29; *see also* portable mills
saws, *see* band saws; circular saws; gang saws; head saws; power saws; trim saws; whipsawing
scaling, 179, 195
Scandinavian logging system, 213, 233
Schmidt, Oscar, 160
Schmidt, R.V., 183
Schneider Logging, 172, 174
Seaboard Lumber Sales, 159
Seafoot, Bill, 203
Seaman, Ken, 198
Selkirk Spruce Mills, 119, 201
Shay, Ephraim, 71
Shelley: logging, 155; sawmilling, 86–91, 101, 172
Shelley Sawmills, 86
Shelter Bay: logging, 230
Sherbinin, George, 200
Sherbinin, John, 198–201
Shields, James, 42
Shields, John, 28, 42
shingle mills, 46
Shuswap and Okanagan Railway, 34
Shuswap area: logging, 42, 50, 53–54
Shuswap Lake, 24, 42
Shuswap Lumber, 121
Shuswap Milling, 19, 22, 27–28
Shuswap River, 54–57
Shuswap Shingle and Lumber, 42
Sicamous, 42
Sigalet, Henry, 81–82, 123–24
Sigalet, Jack, 124
Silverton: logging, 62; sawmilling, 50
Silverton Lumber and Power, 50
Simard, Henry, 54
Simard, Napoleon, 123
Simpson, Horace, 124–25
Simpson, Stanley M., 81, 124–25, 159, 183, 195, 220
Simpson, S.M., Lumber, 34, 81, 124–25, 144, 159, 183, 195, 220
Sims, Morris, 154
Sinclair Spruce Mills (Sinclair Lumber), 84, 87, 89, 95, 101, 116, 130, 221
Skeena Crossing, 104
skidders, 99–100, 101, 134, 186–89, 199, 203, 216, 217, 232
Skookumchuck (Shuswap River), 55, 57
Slater, Jack, 46
sleighs and sloops, 24–25, 37, 66, 69, 89, 111, 161
slide-asses, 86, 87
Slim Creek: logging, 155
Sloan, Gordon, 141
Slocan area: logging, 189, 214; sawmilling, 29, 50, 144, 221
Slocan Forest Products, 116, 128, 195, 201, 216, 221, 224, 229, 232, 225–29
Slocan River, 29, 50, 55
sloops, *see* sleighs
Small Business Forest Enterprise Program, 211, 235
small-log utilization, 219
Smith, S.C., 34, 52
Smith family, 41, 118, 183
Smithers area: logging, 186; sawmilling, 104, 128, 186, 198, 218
Soderburg, Ted, 142
Sommers, Robert, 144
Sovereign Creek: logging, 188
Sovereign Lumber, 42
Special Timber Licences, 51, 61, 62
Spillimacheen River, 110
Sproule Creek: sawmilling, 47
spruce, 12, 37, 83, 84, 90, 97, 122, 133, 143, 147, 183, 219
Spurr, Roy, 83, 84, 91–95, 130
Stanyer, Jack, 106
Staples, Otis, 64–65, 76
Staples Lumber, 74, 100, 101
steamboats, 14, 15, 21, 33, 109
Steele, William H., 122
Steele, William H., Lumber, 183
Stewart, Bob, 221
Stinson, Fred, 151, 178, 228
Stokes, John, 174, 212
Stoney Creek bridge, 23
strikes: *see* unions and organized labour
Strimbold, Archie, 105, 189, 199
Strother, Jim, 183
stumpage, 59, 82, 115, 122, 163–65, 175, 179, 195, 208, 217, 235
Sugar Lake, 81–82
Summerland: sawmilling, 183
Summerland Box, 183
Summit Lake: logging, 62, 162; sawmilling, 50
Swanky, Gordon, 224
Swanson family, 228
Sweet Creek, 154

Tackama Forest Products, 201, 228, 229
Taft: sawmilling, 74
Taghum: logging, 47
Takla: logging, 214; sawmilling, 230
Tanglefoot Creek, 69
Taylor Brothers, 183
Taylor: pulp milling, 229; sawmilling, 221
Telkwa: logging, 107; sawmilling, 128, 186
tenure, 27, 29, 61, 76–79, 141–44, 165–66, 178–79, 209, 211, 215, 231
Terrace area: logging, 104
TF&M (Telkwa), 128
Tharion, Roger, 140
The Pas Lumber, 85, 195, 221
Therrien, Dollard, 140, 155
Therrien, Robert, 155
third-band timber, 195, 202, 219, 220
Thompson, Jack, 150, 158
Thompson River Lumber, 28
Thorlakson family, 218
Thorlakson, Al, 201–2, 219, 220
Thorlakson, Doug, 201
Thorlakson, Harold, 201–2, 218, 219
Thorlakson, John, 201
Three Valley Gap: sawmilling, 74
tie cutting, 21–22, 29, 50, 51, 65, 101, 103–9, 121, 142, 143
timber auctions, 164–65
timber berths, 27, 29
Timber Products Stabilization Act, 208
timber quotas, 165–66, 194, 235
Timber Sale Harvesting Licence, 180
timber supply, 27–28, 29, 40, 45, 51, 53, 58–59, 61–62, 63, 76–79, 81, 83, 97–98, 106, 141, 144, 147, 159, 163–65, 169, 186, 193, 208, 209, 219, 225, 230, 234, 235–36
Timber West, 140
Tolko Industries, 202, 218–19
Torpy Creek: logging, 155
Tranquille, 18
Trautman & Garroway, 124, 125
Tree Farm Licences, 179–80, 200, 211, 215, 218
Triangle Pacific Forest Products, 225, 226, 227
trim saws, 42, 46
Tripac Stud Mill, 225
Trout Lake, 45
truck logging, 119, 122, 125, 134, 143, 186, 202–3, 212, 231, *see also* arch trucks
Tum Tum Lake, 42
Two-by-Four Stud Mill, 140
Two Mile Planing Mills, 150, 155

Umatilla, 14
unions and organized labour, 62, 101, 129–31, 209
United Grain Growers, 83
Uphill, Bill, 122, 206, 207, 233
Upper Arrow Lake, 74
Upper Fraser Mills, 95, 143
Usk: logging, 104

Valhalla Wilderness Society, 214
value-added (remanufacturing), 235
Van Drimmelen, Nick, 174, 201
Vander Zalm, Premier Bill, 215
Vanderhoof: sawmilling, 116, 208, 229
Veerbeek, John, 186, 189, 235
Vernon area: logging, 115; sawmilling, 122–23, 183, 220
Vernon Box & Pine Lumber, 183
Victoria-Yukon Trading, 34

Waldie, Carlos, 160
Waldie, Fred, 230, 234
Waldie, William, 46–47, 111–14
Waldie & Sons, 47, 138, 144
Waldo, 78; sawmilling, 62, 74
Wardner, 13, 78; sawmilling, 63
waste wood, 166, 169, 172–89, 193, 195
water power, 15–16, 17, 18, 19, 27, 103
Waterland, Tom, 212
Watson Lake: sawmilling, 235
Watts, A.E., 64
Wattsburg, *see* Lumberton
Weatherhead, 106
Webster, Dan, 109
Webster, Dave, 212
Weldwood Canada, 139, 154, 201, 216, 224, 231
Wenner-Gren, Axel, 171, 179
West Fraser Timber, 109, 122, 124, 133, 152, 158, 169, 195, 198, 218, 235
Westar, 218, 229
Westbridge: sawmilling, 201
Western Pine Lumber, 183
Western Plywood, 83, 139, 144, 154, 201, 224
western red cedar, 12, 37, 43, 46, 50, 52, 53, 183
Weyerhaeuser, 181, 183, 208, 229
Whatshan Lake, 47
whipsawing (pit sawing), 15
white pine, 37, 46, 82
Whonnock Industries, 125, 201
Wigen, Monrad, 64
Wilder, Earl, 128
Wilder, Lloyd, 128
Wilder Brothers Lumber, 201
Wildhorse, 17
Williams, Bob, 207
Williams Lake area: logging, 139, 207, 231, 232; sawmilling, 140, 141, 155, 158, 166, 168, 197, 198, 218, 220, 234
Williston, Ray, 145, 165, 166, 168, 172, 174, 175, 178, 179, 181, 182, 194, 211, 212, 219
Williston Lake area: logging, 199, 203, 232
Willow River: sawmilling, 83, 105
Willow River Lumber, 105
Wineau: sawmilling, 29
Winfield: sawmilling, 34
Winlaw, J.B., 111
Winther, Chris, 224
Winther, Paul, 224
Winton family, 84–85, 98, 100, 221
Winton, C.J., 85
Winton, David, 85
Winton Lumber, 84
Witherby Tree Farm, 174
Wood, Bob, 212
Wood & Cargill, 41
Wood & McNab Lumber, 62–63
Woodlot Licences, 211
Woods, Angus, 54
Woodworkers Industrial Union of Canada, 130
World War II, 114–19
Wrangler, Joe, 138
Wright, G.B., 17
Wright, Tom, 166, 172, 174
Wycliffe, 64–65, 78; sawmilling, 74
Wynndel: sawmilling, 64, 111
Wynndel Box & Lumber, 64

Yahk, 78; sawmilling, 106
Yale area: logging, 21, 22; sawmilling, 15, 22, 29
Yale-Columbia Lumber, 29
yellow cedar, 201
Yoder, Alec, 50
Young, J., & Sons, 150

Zimmerman, Adam, 174